HANDBOOK OF
CHEMICAL VAPOR DEPOSITION (CVD)

HANDBOOK OF CHEMICAL VAPOR DEPOSITION (CVD)

Principles, Technology and Applications

by

Hugh O. Pierson

Consultant and Sandia National Laboratories (retired)
Albuquerque, New Mexico

NOYES PUBLICATIONS
Westwood, New Jersey, U.S.A.

Library of Congress Catalog Card Number: 91-46658
ISBN: 0-8155-1300-3
Printed in the United States

Published in the United States of America by
Noyes Publications
Fairview Avenue, Westwood, New Jersey 07675

Transferred to Digital Printing, 2011

Printed and bound in Great Britain by
CPI Antony Rowe, Chippenham and Eastbourne

Reprint Edition

Library of Congress Cataloging-in-Publication Data

Pierson, Hugh O.
        Handbook of chemical vapor deposition (CVD) : principles,
technology, and applications / by Hugh O. Pierson.
            p.   cm.
        Includes bibliographical references and index.
        ISBN 0-8155-1300-3
        1. Vapor-plating--Handbooks, manuals, etc. I. Title.
TS695.P52   1992
671.7'35--dc20   92-46658
                                                                      CIP

MATERIALS SCIENCE AND PROCESS TECHNOLOGY SERIES

Editors

Rointan F. Bunshah, University of California, Los Angeles *(Series Editor)*
Gary E. McGuire, Microelectronics Center of North Carolina *(Series Editor)*
Stephen M. Rossnagel, IBM Thomas J. Watson Research Center
(Consulting Editor)

Electronic Materials and Process Technology

DEPOSITION TECHNOLOGIES FOR FILMS AND COATINGS: by Rointan F. Bunshah et al

CHEMICAL VAPOR DEPOSITION FOR MICROELECTRONICS: by Arthur Sherman

SEMICONDUCTOR MATERIALS AND PROCESS TECHNOLOGY HANDBOOK: edited by Gary E. McGuire

HYBRID MICROCIRCUIT TECHNOLOGY HANDBOOK: by James J. Licari and Leonard R. Enlow

HANDBOOK OF THIN FILM DEPOSITION PROCESSES AND TECHNIQUES: edited by Klaus K. Schuegraf

IONIZED-CLUSTER BEAM DEPOSITION AND EPITAXY: by Toshinori Takagi

DIFFUSION PHENOMENA IN THIN FILMS AND MICROELECTRONIC MATERIALS: edited by Devendra Gupta and Paul S. Ho

HANDBOOK OF CONTAMINATION CONTROL IN MICROELECTRONICS: edited by Donald L. Tolliver

HANDBOOK OF ION BEAM PROCESSING TECHNOLOGY: edited by Jerome J. Cuomo, Stephen M. Rossnagel, and Harold R. Kaufman

CHARACTERIZATION OF SEMICONDUCTOR MATERIALS–Volume 1: edited by Gary E. McGuire

HANDBOOK OF PLASMA PROCESSING TECHNOLOGY: edited by Stephen M. Rossnagel, Jerome J. Cuomo, and William D. Westwood

HANDBOOK OF SEMICONDUCTOR SILICON TECHNOLOGY: edited by William C. O'Mara, Robert B. Herring, and Lee P. Hunt

HANDBOOK OF POLYMER COATINGS FOR ELECTRONICS: by James J. Licari and Laura A. Hughes

HANDBOOK OF SPUTTER DEPOSITION TECHNOLOGY: by Kiyotaka Wasa and Shigeru Hayakawa

HANDBOOK OF VLSI MICROLITHOGRAPHY: edited by William B. Glendinning and John N. Helbert

CHEMISTRY OF SUPERCONDUCTOR MATERIALS: edited by Terrell A. Vanderah

CHEMICAL VAPOR DEPOSITION OF TUNGSTEN AND TUNGSTEN SILICIDES: by John E.J. Schmitz

ELECTROCHEMISTRY OF SEMICONDUCTORS AND ELECTRONICS: edited by John McHardy and Frank Ludwig

HANDBOOK OF CHEMICAL VAPOR DEPOSITION: by Hugh O. Pierson

(continued)

v

Ceramic and Other Materials—Processing and Technology

SOL-GEL TECHNOLOGY FOR THIN FILMS, FIBERS, PREFORMS, ELECTRONICS AND SPECIALTY SHAPES: edited by Lisa C. Klein

FIBER REINFORCED CERAMIC COMPOSITES: by K.S. Mazdiyasni

ADVANCED CERAMIC PROCESSING AND TECHNOLOGY—Volume 1: edited by Jon G.P. Binner

FRICTION AND WEAR TRANSITIONS OF MATERIALS: by Peter J. Blau

SHOCK WAVES FOR INDUSTRIAL APPLICATIONS: edited by Lawrence E. Murr

SPECIAL MELTING AND PROCESSING TECHNOLOGIES: edited by G.K. Bhat

CORROSION OF GLASS, CERAMICS AND CERAMIC SUPERCONDUCTORS: edited by David E. Clark and Bruce K. Zoitos

HANDBOOK OF INDUSTRIAL REFRACTORIES TECHNOLOGY: by Stephen C. Carniglia and Gordon L. Barna

Related Titles

ADHESIVES TECHNOLOGY HANDBOOK: by Arthur H. Landrock

HANDBOOK OF THERMOSET PLASTICS: edited by Sidney H. Goodman

SURFACE PREPARATION TECHNIQUES FOR ADHESIVE BONDING: by Raymond F. Wegman

FORMULATING PLASTICS AND ELASTOMERS BY COMPUTER: by Ralph D. Hermansen

HANDBOOK OF ADHESIVE BONDED STRUCTURAL REPAIR: by Raymond F. Wegman and Thomas R. Tullos

FOREWORD

When, in 1946, the writer was introduced to the deposition of coatings by reaction of gaseous precursors at a heated surface, the technique was mainly a laboratory curiosity with few applications, but with much promise. The de Boer-Van Arkel iodide refining process, which revealed the inherent ductility of the Group-IV metals, was elegant in its simplicity, an entire chemical processing plant in a glass bulb. Reactions for depositing silicon, boron, carbon, carbides, nitrides, oxides, borides, silicides, and many other elements and compounds were known, but the potential applications had not yet been realized.

Once disparaged at that time as "Butterfly Chemistry" by a prominent metallurgist, the technique is now the backbone of microelectronic device processing. It has revolutionized fabrication methods throughout industry by providing wear-resistant coatings for cutting, forming, and embossing tools, and for dies, thread guides, and the like. Many aspects of the rapidly emerging composite materials technology depend upon this now-important technique. Other current and potential applications are equally significant.

In 1966, the book "Vapor Deposition" (C.F. Powell et al, John Wiley & Sons, Inc., New York) was published, which attempted to relate the chemistry, thermodynamics, kinetics, flow dynamics and structural aspects of the technique and its products. The distinction between "Chemical Vapor Deposition" (CVD) and "Physical Vapor Deposition" (PVD), introduced earlier, was retained, but with the recognition that the emerging hybrid techniques of plasma-assisted CVD and reactive ion plating were only representative of a broad spectrum of possibilities. That volume served for many years as a guide for those entering the field of CVD.

During the quarter century since that publication, the development of CVD in both understanding and breadth of application has been explosive. Many books have appeared covering individual aspects of that development. However, a sequel covering the wide expanse of CVD technology has been needed. Hugh Pierson has filled that need with the present volume.

Hugh Pierson is best known for his work at Sandia Laboratories with the refractory borides and other binary CVD coatings. However, he has also had the broad experience in CVD necessary to present its many aspects intelligently. He is to be congratulated for having accomplished this formidable task in such a meaningful way.

January 1992   John M. Blocher, Jr.
Oxford, Ohio

PREFACE

Chemical vapor deposition (CVD) has grown very rapidly in the last twenty years and applications of this fabrication process are now key elements in many industrial products, such as semiconductors, optoelectronics, optics, cutting tools, refractory fibers, filters and many others. CVD is no longer a laboratory curiosity but a major technology on par with other major technological disciplines such as electrodeposition, powder metallurgy or conventional ceramic processing.

The estimated market for CVD products is very large and predicted to reach almost three billion dollars in 1993, in the U.S. alone. The rapid development of the technology and the expansion of the market are expected to continue in the foreseeable future.

The reasons for the success of CVD are simple:

- CVD is a relatively uncomplicated and flexible technology which can accommodate many variations.
- With CVD, it is possible to coat almost any shape of almost any size.
- Unlike other thin film techniques such as sputtering, CVD can also be used to produce fibers, monoliths, foams and powders.
- CVD is economically competitive.

Still the technology faces many challenges and a large research and development effort is under way in most major laboratories in the U.S., Japan and Europe, particularly in the semiconductor and tool industries. The general outlook is one of constant change as new designs, new products and new materials are continuously being introduced.

Generally CVD is a captive operation and an integral part of a fabrication process, particularly in microelectronics where most manufacturers have their own CVD facilities. It is also an international business, and, while the U.S. is still in the lead, a great deal of work is under way, mostly in Japan and Europe. A significant trend is the internationalization of the CVD industry, with many multi-national ventures.

Several books have recently been published on the subject of CVD, each dealing with a specific aspect of the technology, such as CVD for microelectronics or metallo-organic CVD. With many of his colleagues, the author has felt the need for a general, systematic, objective and balanced review solely devoted to CVD, which would cover all its scientific, engineering and applications aspects, coordinate the divergent trends found today in the CVD business, promote interaction and sharpen the focus of research and development.

To fill this need is the purpose of this book, which should be useful to students, scientists and engineers, as well as production and marketing managers and suppliers of materials, equipment and services.

The author is fortunate to have the opportunity, as a consultant, to review and study CVD processes, equipment, materials and applications for a wide cross-section of the industry, in the fields of optics, optoelectronics, metallurgy and others. He is in a position to retain an overall viewpoint difficult to obtain otherwise.

The book is divided into three major parts. The first covers a theoretical examination of the CVD process, a description of the major chemical reactions and a review of the CVD systems and equipment used in research and production, including the advanced sub-processes such as plasma, laser, and photon CVD.

The second part is a review of the materials deposited by CVD, i.e., metals, non-metallic elements, ceramics and semiconductors, and the reactions used in their deposition.

The third part identifies and describes the present and potential applications of CVD in semiconductors and electronics, in optics and optoelectronics, in the coating of tools, bearings and other wear- and corrosion-resistant products, and in the automobile, aerospace and other major industries.

The author is indebted to Dr. Jacob Stiglich and other members of the staff at Ultramet for their cooperation, and to an old friend, Dr. John M. Blocher Jr., for his many ideas, comments and thorough review of the manuscript. He is also grateful to George Narita, Executive Editor of Noyes Publications, for his help and patience in the preparation of this book.

February 1992   Hugh Pierson
Albuquerque, New Mexico

NOTICE

To the best of our knowledge the information in this publication is accurate; however, the Publisher does not assume any responsibility or liability for the accuracy or completeness of, or consequences arising from, such information. Mention of trade names or commercial products does not constitute endorsement or recommendation for use by the Publisher.

Final determination of the suitability of any information or product for use contemplated by any user, and the manner of that use, is the sole responsibility of the user. We recommend that anyone intending to rely on any recommendation of materials or procedures for chemical vapor deposition mentioned in this publication should satisfy himself as to such suitability, and that he can meet all applicable safety and health standards. We strongly recommend that users seek and adhere to the manufacturer's or supplier's current instructions for handling each material they use.

CONTENTS

xiii

1

INTRODUCTION AND GENERAL CONSIDERATIONS

1.0 INTRODUCTION

Chemical vapor deposition (CVD) is a very versatile process used in the production of coatings, powders, fibers and monolithic components. With CVD, it is possible to produce almost any metal and non-metallic element, including carbon or silicon, as well as compounds such as carbides, nitrides, oxides, intermetallics and many others. This technology is now an essential factor in the manufacture of semiconductors and other electronic components, in the coating of tools, bearings and other wear resistant parts and in many optical, opto-electronic and corrosion applications. The market for CVD products is already very large both in the US and abroad and is expected to reach several billions dollars by the end of the century.

The very wide range of CVD products is illustrated by the following recent commercial products (1991):

   • A speaker diaphragm with improved acoustical properties coated with a thin film of diamond obtained by plasma CVD

• Diamond-like carbon coatings produced by plasma CVD for bushings and textile components with much improved wear resistance

• Titanium carbide and titanium nitride CVD-coated carbide tools which greatly outperform uncoated tools and are rapidly taking over the industry

• Iridium, deposited by metallo-organic CVD, which has shown remarkable resistance to corrosion in small rocket nozzles at temperatures up to 2000°C

• Tungsten, silicon oxide, metal silicides and other coatings produced by CVD which now constitutes the dominant process in the production of advanced semiconductor components

• Energy saving optical coatings for architectural glass by atmospheric pressure CVD, produced *in situ* during the processing of float glass

• Pyrolytic boron nitride crucibles produced by CVD with outstanding chemical inertness, which are used extensively in the electronic industry

• CVD boron fibers which are extremely stiff and strong and are used as reinforcement in structural components of USAF fighter planes

Chemical vapor deposition may be defined as the deposition of a solid on a heated surface from a chemical reaction in the vapor phase. It belongs to the class of vapor transfer processes which are atomistic in nature; that is, the deposition species are atoms or molecules or a combination of these. Beside CVD, they also include the physical vapor deposition processes (PVD) such as evaporation, sputtering, molecular beam epitaxy, ion plating and ion implantation as well as pack cementation which is a hybrid process combining vapor phase transfer and solid state diffusion.

In many respects, CVD competes directly with these processes, but it is also used in conjunction with them and many of the newer processes are actually hybrids of the two systems such as plasma enhanced CVD or activated sputtering. A general knowledge of the PVD processes is important for the full understanding and appreciation of CVD and a summary description is given in the appendix.

CVD has several important advantages which make it the preferred process in many cases. These can be summarized as follows:

• As generally used at pressures above the molecular flow region, it is not restricted to a line of sight deposition which is generally characteristic of sputtering, evaporation and other PVD processes. As such, CVD has high throwing power. Deep recesses, holes and other difficult three-dimensional configurations can usually be coated with relative ease. For instance, integrated circuit via holes with an aspect ratio of 10/1 can be completely filled with CVD tungsten.

• The deposition rate is high and thick coatings can be readily obtained (in some cases centimeters thick) and the process is generally competitive and, in many cases, more economical than the PVD processes.

• The CVD equipment is relatively simple, does not require ultra-high vacuum, and generally can be adapted to many process variations. Its flexibility is such that it allows many changes in composition during deposition and codeposition of compounds is readily achieved.

CVD however is not the universal coating panacea. It has several disadvantages, a major one being that it is most versatile at temperatures of 600°C and above where the thermal stability of the substrate may limit its applicability. However the development of plasma CVD and metallo-organic CVD partially offsets this problem. Another disadvantage is the requirement of having chemical precursors (the starter materials) with high vapor pressure which are often hazardous and at times extremely toxic. The by-products of these precursors are also toxic and corrosive and must be neutralized, which may be a costly operation.

2.0 HISTORICAL PERSPECTIVE

CVD is not a new process. As reviewed in the pioneer work of Powell, Oxley and Blocher (1), its first practical use was developed in the 1880's in the production of incandescent lamps to improve the strength of filaments by carbon or metal deposition (1). In the same decade, the carbonyl nickel process was developed by Ludwig Mond and others. A number of patents were issued during that period covering the basis of CVD (2).

The process developed slowly in the next fifty years and was limited mostly to pyro- and extraction metallurgy for the production of high purity refractory metals such as tantalum, titanium and zirconium, using the carbonyl reaction (the Mond process), the iodide decomposition reaction (the de Boer-Van Arkel process) or the magnesium reduction reaction (the Kroll process). These reactions are described in Chapters Three and Five of this present book.

It is only since the end of World War II that the process began to be used on a large scale as researchers realized the potential of deposition and the formation of coatings and free-standing shapes in addition to metallurgical extraction. Its importance has been growing ever since.

Other important dates in the development of CVD are the following:

- 1960: introduction of the terms "CVD" and "PVD" to distinguish "chemical vapor deposition" from "physical vapor deposition" (3)

- 1960: introduction of CVD in semiconductor fabrication

- 1960: concept of CVD TiC coating of cemented carbide tools introduced (4)

- 1963: basis of plasma CVD in electronics established (5)

- 1968: start of industrial use of CVD coated cemented carbides

- 1968: first development of metallo-organic CVD for the deposition of gallium arsenide

- 1974: introduction of plasma-enhanced CVD

- 1976: first applications of low pressure CVD in electronics

- 1983: introduction of electron cyclotron resonance for plasma CVD

Today, the technology is developing at an increasingly rapid rate. However, to retain the right perspective, it should be noted that it took over a century of steady and continuous scientific and engineering efforts to reach the present state of the art. Yet, in spite of this progress, many formidable challenges remain, such as the accurate prediction of a given CVD composition and its structure and properties. In fact, even though the understanding of the theory and mechanism of CVD has made great progress, CVD itself can still be considered as much an art as a science and progress continues to rely for a very large part on experimental developments.

3.0 THE APPLICATIONS OF CVD

CVD is a very versatile and dynamic technology which is constantly expanding and improving as witnessed by the recent developments in metallo-organic CVD, plasma CVD, laser CVD and many others. As the technology is expanding, so is the scope of its applications. This expansion is the direct result of a large research effort carried out by many workers in industry, the universities and government laboratories. The outlook is one of constant changes as new designs, new products and new materials are continuously being introduced; one process will grow at the expense of another to be replaced in time by another improved one.

Two major areas of application of CVD have rapidly developed in the last twenty years or so, namely in the semiconductor industry and in the so-called metallurgical coating industry which is presently dominated by cutting tool coatings. Progress in these two areas seems to occur independently with little interaction when actually they share the same scientific basis, the same principles, the same chemistry and in many cases the same equipment. A purpose of this book is to bring these divergent areas together in one unified whole and to accomplish, in a book form, what has been the focus of the International CVD and the Euro-CVD conferences.

CVD applications can be classified by product functions such as electrical, opto-electrical, optical, mechanical and chemical. This classification corresponds roughly to the various segments of industry such as the electronic industry, the optical industry, the tool industry and the chemical industry. CVD applications can also be classified by product form such as coatings, powders, fibers, monoliths and composites.

Inevitably, there is a certain degree of overlapping between these two general classifications. For instance, optical applications are found in both coating and fiber functions and fibers are used in optics and in structural and mechanical applications. These relationships will be reviewed in the chapters on the applications of CVD.

4.0 CVD COATINGS

Coatings are by far the largest area of application of CVD at the present but by no means the only one and the applications for powders, fibers, monoliths and composites are growing rapidly.

Coatings are used on a large scale in many production applications in optics, electronics, opto-electronics, tools, wear and erosion and others. In the case of electronics and opto-electronics practically all CVD applications are in the form of coatings.

4.1 Composite Nature of Coatings

The surface of a material exposed to the environment experiences wear, corrosion, radiation, electrical or magnetic fields and other phenomena. It must have the properties needed to withstand the environment or to provide certain desirable properties such as reflectivity, semi-conductivity, high thermal conductivity or erosion resistance. Depositing a coating on a substrate produces a composite material and, as such, allows it to have surface properties which can be very different from those of the bulk material (6).

In some cases, a coating may be very similar to the substrate and CVD can control the subtle changes necessary to provide the desired result. An example is the multi-layered deposition of semi-conductor material in the so-called strained layer superlattice, a composite which could possibly be the building block of the next generation of supercomputers. Table 1 summarizes the surface properties that can be obtained or modified by the use of coatings.

TABLE 1

Material Properties Affected by Coatings

Electrical	Resistivity
	Superconductivity
	Crystal lattice control
	Magnetism
	Dielectric constant
Optical	Refraction
	Emissivity
	Reflectivity
	Photoconductivity
	Selective absorption
Mechanical	Wear
	Friction
	Hardness
	Adhesion
	Toughness
	Ductility
	Strength

TABLE 1 (cont.)

Material Properties Affected by Coatings

Chemical	Diffusion
	Corrosion
	Oxidation
	Catalysis
	Electrochemical
Porosity	Surface area
	Pore size
	Pore volume

4.2 Major Coating Processes

The last decade has seen a proliferation of coating processes and a very rapid evolution of the technology. The major coating processes currently in use are listed in Table 2.

TABLE 2

Classification of Major Deposition Processes

Liquid	Sol-gel
	Paint
	Dipping
	Liquid phase epitaxy
	Electrophoresis

TABLE 2 (cont.)

Classification of Major Deposition Processes

Electrolytic	Electroplating
	Electroless plating
	Fused salt electrolysis
Particulate	Plasma spraying
	D-gun spraying
	Flame spraying
	Electrophoresis
Physical Vapor	Sputtering
Deposition	Evaporation
(Thin Film)	Ion plating
	Ion implantation
	Molecular beam epitaxy
Chemical Vapor	
Deposition	

Liquid based coatings are a well known and very large industry. With the exception of sol-gel, they are not suitable in applications requiring very thin and closely controlled films.

Electrolytic coatings have been used for decades. They are inexpensive and have generally good properties. Porosity is a problem and their use in thin film applications requiring high density or optical quality is generally not possible. The process may also cause severe environmental pollution.

Particulate deposition, such as plasma spraying, is another category of coating which is relatively inexpensive and widely used in corrosion- and wear-control applications. However it cannot normally produce very thin, high quality films such as required in optics and electronics.

5.0 PROFILE OF THE CVD INDUSTRY

This book is primarily directed toward the study of the CVD technology. However, it is important to consider this technology within the more general context of the CVD industry and a brief review of the industry is necessary in order to bring the technology into proper focus.

An overview of the CVD market classified by function applications is given in Table 3.

Each of these function applications will be reviewed in detail in subsequent chapters. These figures are at best an estimate since the precise share of CVD, in terms of cost or add-on value, is often difficult to isolate from the overall production cost. For instance, the fabrication of semiconductor chips is a complicated and lengthy procedure which involves many steps, including lithography, cleaning, etching, testing and oxidation not to mention single crystal silicon growth and preparation. CVD is an integral part of this fabrication sequence and is used extensively (and in some cases exclusively) in the production of insulating, conductive and semiconductor films. But to isolate its exact cost is difficult. In addition, this cost may vary considerably from one application to another and from one manufacturer to another. For these reasons, only a rough estimate can be made.

TABLE 3

Overview of the Estimated CVD Market by Function Applications (US only)

	Annual Growth Rate%	1988	1993 Projection
		($ million)	
Electronics	20	$940	$2330
Optoelectronics	20	42	140
Photovoltaic	15	17	34
Optical	25	55	167
Structural	10	145	233
Chemical	10	12	19
TOTAL		$1211	$2923

(Source: G.A.M.I., Gorham, ME 04038)

As can be seen, in 1988 the electronic applications formed a large majority of the applications of CVD (77%), followed by structural and mechanical applications (12%), the latter being mostly cutting tools. These percentages are not expected to change significantly in the next few years.

CVD production, both in the US and abroad, is essentially captive; that is, the CVD processing is an integral part of sequential operations. This is particularly true in the field of semiconductor and microelectronics where most manufacturers have their own CVD facilities which, more often than not, are incorporated within a production line. Likewise, the manufacturers of cutting tools, turbine blades and other components that require hard coatings also have their own CVD facilities. There are few independent CVD producers as such and these usually specialize in specific applications or rely on their expertise for research and development of new processes and materials.

CVD is now truly an international business and no longer a US monopoly. In Japan, in other Pacific Rim countries and in Europe, both Western and Eastern, a great deal of work is under way, in research as well as in production. The US is still in the lead with an estimated 40% of the business, followed by both Europe and Japan with 25% each, the balance being for the rest of the world.

The amount of research and development in CVD is considerable not only in industry, but also in universities and in all the major government laboratories, in the US as well as abroad. Government sponsorship in most countries remains a very major factor either in terms of funding or as a research leader and organizations, such as the Department of Defense (DOD), the Department of Energy (DOE) and NASA in the US, MITI and NIRIM in Japan, CNRS in France and many others, play leading roles in the development of CVD.

Until about 1980, CVD equipment was designed and built in-house and a CVD equipment industry essentially did not exist as such. Since then, there has been a considerable shift to standardized systems built by specialized equipment manufacturers. This is the result of increasing sophistication and cost of the technology, particularly in semiconductor and microelectronic fabrication. The CVD equipment market is now of considerable size and is attracting a large number of companies. It was estimated to reach one billion dollars worldwide in 1990. The business is still led by the US with an estimated share of slightly less than 50% in 1988. These figures actually become less relevant with the rapid internationalization of the industry. Large foreign companies now have branches and plants in the US and many major US producers maintain operations abroad. This appears to be the shape of the future.

6.0 TRENDS IN CVD

A review of the proceedings of the various international conferences on CVD, such as the ones sponsored by the Electrochemical Society since the 1960's and the more recent European conferences, provides an accurate cross section of the past trends in the development of CVD. In the 1960's, emphasis was on the CVD of refractory metals particularly tungsten. This was followed by a gradual shift to other refractory metals and composites. In the 1970's, the development of CVD for semi-conductor/microelectronic applications became the dominant interest with primary stress on silicon. The development of machine-tool coatings was also important in this decade. In the 1980's, the interest broadened to cover a multitude of materials and processes including metallo-organic CVD (MOCVD), plasma CVD, laser CVD and many others.

Looking at the future now, it is safe to predict the ever increasing role of CVD in the semiconductor-microelectronics and opto-electronics areas where CVD is already a major contributor. In other fields such as optics, electro-magnetics and ferroelectrics, CVD is now in the early stage of development but will undoubtedly see a marked upsurge often at the expense of other vapor deposition processes. This is due in part to its excellent throwing power and ability to coat three dimensional surfaces, features which cannot be matched by the physical vapor deposition processes.

MOCVD, plasma-CVD and photo-CVD will see a rapid development, not only in the semiconductor-microelectronic area but also in hard coating, erosion and wear applications since their lower deposition temperatures now permit the use of a broader spectrum of substrates.

Diamond and diamond-like coatings, while still in the R&D stage, are on the threshold of a rapid expansion, especially if the problems of epitaxy and single crystal growth can be solved satisfactorily.

Advances in CVD equipment design will be centered in two major areas.

• the integration of various processes in one piece of equipment, such as CVD, etching, ion implantation and other physical vapor deposition (PVD) processes

• the control of the deposition parameters through in situ observations and the use of basic chemical engineering equipment design

Chemical vapor infiltration (CVI) will become a major process in the production of fiber reinforced ceramic and metal composites and fibers and powders produced or coated by CVD will play a major role in these developments.

7.0 BOOK OBJECTIVES

The objectives of this book are to:

• Review the theoretical aspects of CVD, i.e. chemical thermodynamics, kinetics and gas dynamics

• Provide a detailed assessment of the technology of CVD and its relation to the production of coatings, fibers, powders and monolithic shapes

• Describe the various processes and equipment used in R&D and production such as thermal CVD, plasma CVD, photo CVD, MOCVD and others

• Review the materials that can be produced by CVD

• Identify and describe present and potential CVD applications

The book is divided into three major sections. The first deals with theory and processes, the second with the materials that can be produced by CVD and the third with the present and potential CVD applications. These sections are cross referenced to facilitate review and reduce duplication.

REFERENCES

1. Powell, C.F, Chemical Vapor Deposition, in *Vapor Deposition* (C.F. Powell, J.H. Oxley and J.M. Blocher Jr., Eds.) pp. 249-276, John Wiley & Sons, New York (1966)

2. Sawyer, W.E. and Man, A., US Pat. 229335 (June 29 1880) on pyrolytic carbon; Aylsworth, J.W., US Pat. 553296 (Jan.21 1896) on metal deposition; deLodyguine, A., US Pat. 575002 (Jan.12 1897) and 575668 (Jan.19 1897) on metal deposition; Mond, L., US Pat. 455230 (June 30 1891) on nickel deposition

3. Blocher, J. M. Jr., *J. Electrochem. Soc.*, 107:177C (1960)

4. Ruppert, W., U. S. Patent 2962388

5. Atl, L.L. et al, *J. Electrochemical Soc.*, 110:456 (1963) and 111:120 (1964)

6. Picreaux, S. and Pope, L., Tailored Surface Modifications by Ion Implantation and Laser Treatment, *Science*, 226: 615-622 (1986)

2

FUNDAMENTALS OF CHEMICAL
VAPOR DEPOSITION

1.0 INTRODUCTION

Chemical vapor deposition is a synthesis process in which the chemical constituents react in the vapor phase near or on a heated substrate to form a solid deposit. It may combines several scientific and engineering disciplines including thermodynamics, plasma physics, kinetics, fluid dynamics and of course chemistry. In this chapter, the fundamental aspects of these disciplines and their interconnection will be examined as they relate to CVD.

The chemical reactions used in CVD are numerous and include thermal decomposition (pyrolysis), reduction, hydrolysis, disproportionation, oxidation, carburization and nitridization. They can be used either singly or in combination. They are reviewed in Chapter Three.

These reactions can be activated by several methods which are reviewed in Chapter Four. The most important are as follows:

• thermal activation which typically takes place at high temperature, i.e. >900°C, although the temperature can also be lowered considerably if metallo-organic precursors are used (MOCVD)

• plasma activation which typically takes place at much lower temperature, i.e. 300-500°C

• photo enhanced activation using electromagnetic radiation usually short-wave ultraviolet, which can occur by the direct activation of a reactant or by the activation of an intermediate which then acts as a reacting agent

Until fairly recently, most CVD operations were relatively simple and could be readily optimized experimentally by changing the reaction, the activation method or the deposition variables until a satisfactory deposit was achieved. It is still possible to do just that and, indeed, in some cases it is the most efficient way to proceed. However, many of the CVD processes are becoming increasingly complicated with many more variables which would make the empirical approach far too cumbersome.

A theoretical analysis is, in most cases, an essential step which, if properly carried out, should predict what will happen to the reaction, what the resulting composition of the deposit might be (i.e. stoichiometry), what type of structure to expect (i.e. the geometric arrangement of its atoms) and what the reaction mechanism (i.e. the path of the reaction as it forms the deposit) is likely to be. This analysis may then provide a guideline for an experimental program, considerably reduce its scope and save a great deal of time and effort. However, due to the complexity of the CVD phenomena, a complete and accurate modeling of a CVD reaction is, at this stage, still beyond reach.

To do this analysis, a clear understanding of the CVD process is necessary and a review of several fundamental considerations in the disciplines of thermodynamics, kinetics and chemistry is in order. It is not the intent here to dwell in detail on these considerations but rather provide an overview which should be adequate for the general reader. More detailed investigations of the theoretical aspects of CVD are given in references 1, 2, 3, and 4.

2.0 THERMODYNAMICS OF CVD

A CVD reaction is governed by thermodynamics, that is the driving force which indicates the direction the reaction is going to proceed (if at all), and by kinetics, which defines the transport process and determines the rate control mechanism, in other words, how fast it is going.

Chemical thermodynamics is concerned with the inter-relation of various forms of energy and the transfer of energy from one chemical system to another in accordance with the first and second laws of thermodynamics. In the case of CVD this transfer occurs when the gaseous compounds, introduced in the deposition chamber, react to form the solid deposit (and by-products gases).

2.1 ΔG Calculations and Reaction Feasibility

The first obvious step in any theoretical analysis is to make sure that the desired CVD reaction will take place. This will happen if the thermodynamics is favorable, that is if the transfer of energy (i.e. the free energy change of the reaction known as ΔG_r) is negative. To calculate ΔG_r, it is necessary to know the thermo-dynamic properties of each component, specifically their free energies of formation (also known as Gibbs free energy), ΔG_f. The relationship is expressed as follows:

Eq.(1) $\Delta G_r^0 = \Sigma\ \Delta G_f^0$ products $- \Sigma\ \Delta G_f^0$ reactants

The free energy of formation is not a fixed value but varies as a function of several parameters which include the type of reactants, the molar ratio of these reactants, the process temperature, and the process pressure. This relationship is represented by the following equation:

Eq.(2) $\Delta G_r = \Delta G_f^0 + RT \ln Q$

where: $\Delta G_f^0 = \Sigma z_i^i\ \Delta G_{f,i}^0$

z_i	=	stoichiometric coefficient of species "i" in the CVD reaction (negative for reactants, positive for products)
$\Delta G_{f,i}^0$	=	standard free energy of formation of species "i" at temperature T and 1 atm.
R	=	gas constant
T	=	absolute temperature
Q	=	$\Pi_i a_i^{z_i}$
a_i	=	activity of species "i" which is = 1 for pure solids and = $P_i = x_i P_T$ for gases
P_i	=	partial pressure of species "i"
x_i	=	mole fraction of species "i"
P_T	-	total pressure

By definition, the free energy change for a reaction at equilibrium is zero, hence:

Eq.(3) $\Delta G = - RT \ln K$

where $Q_{eq} = K =$ equilibrium constant

It is the equilibrium conditions of composition and activities (partial pressure for gases) that are calculated to assess the yield of a desired reaction.

The first step is to demonstrate that the reaction is feasible. It is illustrated in the following example regarding the formation of titanium diboride using either diborane or boron trichloride as a boron source as shown in the following reactions:

[1] $TiCl_4 + 2BCl_3 + 5H_2 \longrightarrow TiB_2 + 10HCl$

The changes in free energy of formation of reaction [1] is shown in Figure 1 as a function of temperature (5). The values of ΔG_r were calculated using equation (1) above for each temperature. The Gibbs free energy values of the reactants and products were obtained from the JANAF Tables (6). Other sources of thermodynamic data are listed in reference (7). These sources are generally accurate and satisfactory for the thermodynamic calculations of most CVD reactions; they are constantly being revised and expanded.

TiB_2 can also be obtained using diborane as a boron source as follows:

[2] $TiCl_4 + B_2H_6 \longrightarrow TiB_2 + 4HCl + H_2$

The changes in the free energy of formation on this reaction are shown in Figure 1. It should be noted that the negative free energy change is a valid criterion for the favorability of a reaction only if the reaction as written contains the major species that exist at equilibrium. In the case of reaction [2], it is possible that B_2H_6 has already decomposed to B and H and the equilibrium of the reaction might be closer to:

[2a] $TiCl_4 + 2B + 3H_2 \longrightarrow TiB_2 + 4HCl + H_2$ ($\Delta G = -11.293$)

As can be seen in Figure 1, if the temperature is raised sufficiently, ΔG_R becomes negative and the reaction proceeds. This occurs at much lower temperature with diborane than it does with boron trichloride.

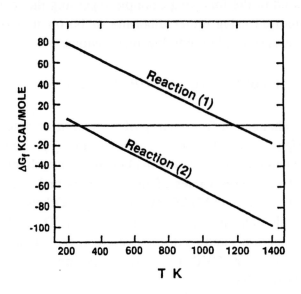

Figure 1. Changes in Free Energy of Formation for TiB$_2$ Deposition in the Following Reactions:

(1) TiCl$_4$ + 2BCl$_3$ $\longrightarrow$ TiB$_2$ + 10HCl

(2) TiCl$_4$ + B$_2$H$_6$ $\longrightarrow$ TiB$_2$ + H$_2$ + 4HCl

2.2 Thermodynamic Equilibrium and Computer Programs

Reactions (1) and (2) above are actually greatly simplified. In reality, it is likely that subchlorides such as TiCl$_3$ and TiCl$_2$ will be formed in reaction (1) and higher boranes in reaction (2). In addition, experimentation shows that the best, fully dense and homogeneous deposits are produced at an optimum negative value of ΔG. For smaller negative values, the reaction rate is very slow and for higher negative values, vapor phase precipitation can occur with resulting sooting. Such factors are not revealed in the simple free energy change calculation.

In many cases, a better understanding of these reactions and a better prediction of the results are needed and a more complete thermodynamic and kinetic investigation is necessary. This is accomplished by the calculation of the thermodynamic equilibrium of a CVD system, which will provide useful information on the characteristics and behavior of the reaction including the optimum range of deposition conditions. It is based on the rule of thermodynamics which states that a system will be in equilibrium when the Gibbs free energy is at a minimum. The objective then is the minimization of the total free energy of the system and the calculation of equilibria at constant temperature and volume or pressure (8). It is a complicated and lengthy calculation but, fortunately, computer programs are now available that considerably simplify the task (9).

Such programs include SOLGASMIX which was developed by Erikson and Besmann (10,11) and EKVICALC and EKVIBASE, developed by Nolang (12). These programs are now used widely in equilibrium calculations in CVD systems.

To operate the program, it is first necessary to identify all the possible chemical species, whether gaseous or condensed phases, that might be found in a given reaction. The relevant thermodynamic properties of these phases are then entered in the program as input data. If properly performed, these calculations will provide the following information:

• the composition and amount of deposited material that is theoretically possible under any given set of deposition conditions, that is at a given temperature, a given pressure and given input concentration of reactant

• the existence of gaseous species and their equilibrium partial pressures

• the possibility of multiple reactions and the number and composition of possible solid phases, with the inclusion of the substrate as a possible reactant

• the likelihood of reaction between the substrate and the gaseous or solid species

All of this is very valuable information which can be of great help. Yet, it must be treated with caution since, in spite of all the progress in thermodynamic analysis, the complexity of many CVD reactions is such that predictions based on thermodynamic calculations are still subject to uncertainty. As stated above, these calculations are based on chemical equilibrium which is rarely attained in CVD reactions.

It follows that, in order to provide a reliable and balanced investigation, it is preferable to combine the theoretical calculations with an experimental program and, hopefully, they will correlate. Fortunately, laboratory CVD experiments are relatively easy to design and carry out; they do not require expensive equipment and results can usually be obtained quickly and reliably.

A classical example, combining theoretical study and laboratory experiments, is the deposition of niobium, originally described by Blocher (13). The following reaction was used:

$$NbCl_5 \ (g) \longrightarrow Nb \ (s) + 2\text{-}1/2Cl_2 \ (g)$$

In Figure 2, the critical deposition temperature of $NbCl_5$ as a function of its initial pressure is shown from experimental data from Blocher (13) and the author. There are two temperature- pressure regions which are separated by a straight line. The metal is deposited only in the region below the line. Above, there is no deposition. The line is a least square fit of the data. Its position was confirmed using the SOLGASMIX computer program.

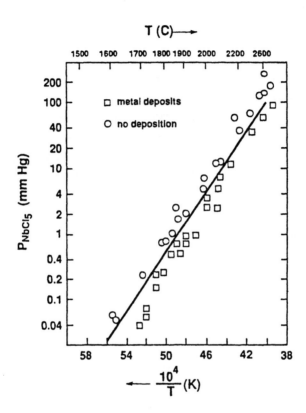

Figure 2. Critical Deposition Temperature of Niobium as a Function of $NbCl_5$ Initial Pressure

This example shows the great degree of flexibility that can be obtained in CVD if a proper understanding of the thermo-dynamics and kinetics is gained. In this particular case, it was pos-sible to deposit a very uniform layer of NbC on a graphite rod, simply by limiting the reaction to the deposition of the carbide. Since the carbide could only be formed using the substrate as a carbon source, the rate was controlled by the diffusion rate of the carbon through the coating and deposition uniformity was achieved over the length of the graphite rod.

3.0 KINETICS AND MASS TRANSPORT MECHANISMS

As shown above, a thermodynamic analysis will indicate what to expect from the reactants as they reach the deposition surface at a given temperature. The question now is, how do these reactants reach that deposition surface? In other words, what is the mass transport mechanism? The answer to this question is very important since the phenomena involved determine the reaction rate and the design and optimization of the CVD reactor.

It should be first realized that a normal CVD process is subject to very complicated fluid dynamics. The fluid (in this case a combination of gases), is forced through pipes, valves and various chambers and, at the same time, is the object of large variations in temperature and to a lesser degree of pressure, before it comes in contact with the substrate and the reaction takes place. The reaction is heterogeneous which means that it involves a change of state, in this case from gaseous to solid. In some cases, the reaction may take place before the substrate is reached while still in the gas phase (gas phase precipitation) as will be reviewed later. As can be expected, the mathematical modeling of these phenomena can be a very complicated task indeed.

3.1 Deposition Sequence

The sequence of events during a CVD reaction is shown graphically in Figure 3 and can be summarized as follows (1):

a) The reactant gases enter the reactor by forced flow.

b) They diffuse through the boundary layer.

c) They are adsorbed on the surface of the substrate where the chemical reaction takes place at the interface; other events such as lattice incorporation and surface motion may also take place at this stage.

d) The gaseous by-products of the reaction are desorbed and diffuse away from the surface, through the boundary layer.

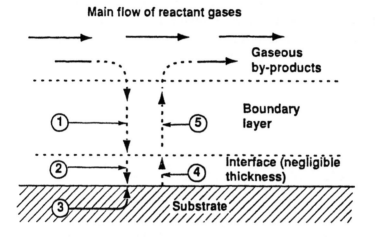

1. Diffusion in of reactants through boundary layer
2. Adsorption of reactants on substrate
3. Chemical reaction takes place
4. Desorption of adsorbed species
5. Diffusion out of by-products

Figure 3. Sequence of Events During Deposition

These steps occur in the sequence shown and the slowest step will determine the rate. The concept of boundary layer applies in most CVD depositions in the viscous flow range where pressure is relatively high. In cases where very low pressure is used (i.e. in the mTorr range), the concept is no longer applicable.

3.2 Deposition in a CVD Flow Reactor

The sequence of events described above occurs at any given spot in a CVD flow reactor. As an example, one can consider a very common reaction which is the deposition of tungsten using the hydrogen reduction of the fluoride as follows:

$$WF_6 + 3H_2 \longrightarrow W + 6HF$$

One can imagine a graphite tube that has to be coated internally with tungsten. As shown schematically in Figure 4(a), the reactant gases are introduced in the upstream side, then flow down the reactor tube and exhaust downstream through the vacuum pump.

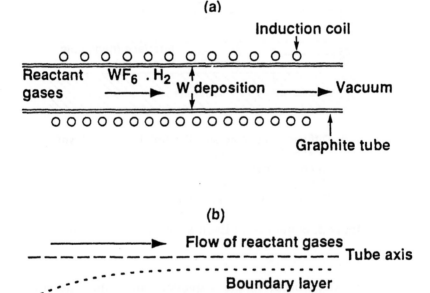

Figure 4. (a) Tungsten Deposition in a Tubular Reactor, (b) Boundary Layer Conditions

The Boundary Layer. What is happening to the gas flow as it enters the tube? This is a problem of fluid mechanics and a complete investigation is outside the scope of this study. It is enough to say that the Reynolds number R_e, which is a dimensionless parameter that characterizes the flow of a fluid, is such that the gas flow is generally laminar, at least in standard CVD reactors, although in some instances the laminar flow may be disturbed by convection gas motion and becomes turbulent. With laminar flow, the velocity of the gas at the deposition surface (the inner wall of the tube) is zero. The boundary is that region in which the flow velocity changes from zero at the wall to essentially that of the bulk gas away from the wall. This boundary layer starts at the inlet of the tube and increases in thickness until the flow becomes stabilized as shown in Figure 4(b). The reactant gases flowing above the boundary layer have to diffuse through this layer to reach the deposition surface as is shown in Figure 3.

The thickness of the boundary layer, D, is inversely proportional to the square root of the Reynolds number as follows:

Eq. (4) $\Delta = \sqrt{\dfrac{x}{R_e}}$

where: $R_e = \dfrac{\rho u_x}{\mu}$

ρ = mass density
u = flow density
x = distance from inlet in flow direction
μ = viscosity

This means that the thickness of the boundary layer increases with lowered gas flow velocity and with the distance from the tube inlet (14).

Under such conditions, what is happening to the velocity of the gases, to the temperature and to the reactant concentrations?

Velocity and Pressure. It is possible to obtain an approximate visualization of the flow pattern by using TiO$_2$ smoke (generated when titanium chloride comes in contact with moist air), although thermal diffusion may keep the smoke particles away from the hot surface where a steep temperature gradient exists. Figure 5 shows a typical velocity pattern in a horizontal tube. As mentioned above, a steep velocity gradient is noticeable going from maximum velocity in the center of the tube to zero velocity at the surface of the wall. The gradient is also shallow at the entrance of the tube and increases gradually downstream.

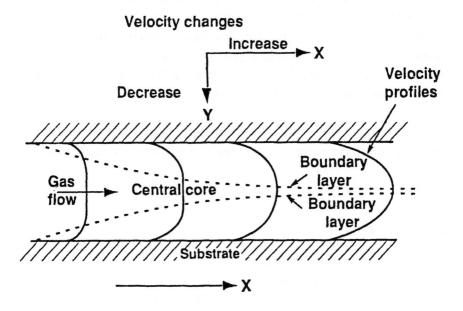

Figure 5. Boundary Layer and Velocity Changes in a Tube Reactor, Showing the Graphs of Velocity Imposed at Different Positions on the Tube

Temperature. Figure 6 shows a typical temperature profile (2). The temperature boundary layer is similar to the velocity layer. The flowing gases heat very rapidly as they come in contact with the hot surface of the tube resulting in a very steep temperature gradient. Overall temperature increases toward downstream.

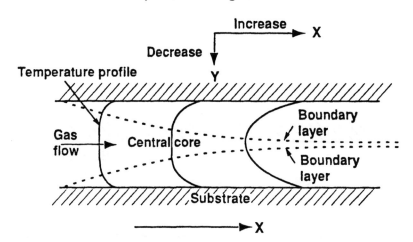

Figure 6. Temperature Boundary Layer and Temperature Changes in a Tubular Reactor, Showing the Graphs of Temperature Imposed at Different Positions on the Tube

Reactant Gas Concentration. As the gases flow down the tube, they become gradually depleted as tungsten is deposited and the amount of HF increases in the boundary layer. This means that, at some point downstream, deposition will cease altogether. The reactant concentration is illustrated in Figure 7. The boundary layers for these three variables (gas velocity, temperature and concentration) may sometimes coincide, although in slow reactions, the profiles of velocity and temperature may be fully developed early on while the deposition reaction is spread well down the tube.

As can be seen, conditions in a flowing reactor, even the simplest such as a tube, may be far from the thermodynamic equilibrium conditions predicted by the equilibrium computer programs. However, in the diffusion controlled range, it is possible to use as the driving force for diffusion the difference between an assumed equilibrium composition at the wall and the bulk gas compositions in the feed (adjusted for downstream depletion) to model some systems to a first approximation.

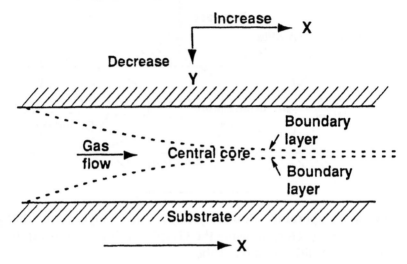

Figure 7. Changes in Reactant Concentration in a Tubular Reactor

3.3 Rate Limiting Steps

What is the rate limiting step of a CVD reaction? In other words, what factor controls the growth rate of the deposit? The answer to this question is critical, since it will help to optimize the deposition reaction, obtain the fastest growth rate and, to some degree, control the nature of the deposit.

This rate limiting step can be either determined by: a) the surface reaction kinetics, b) the mass transport, c) gas-phase kinetics (a more uncommon occurrence).

In the case of control by surface reaction kinetics, the rate is dependent on the amount of reactant gases available. As an example, one can visualize a CVD system where the temperature and the pressure are low. This means that the reaction occurs slowly because of the low temperature and there is a surplus of reactants at the surface since, because of the low pressure, the boundary layer is thin, the diffusion coefficients are large and the reactants reach the deposition surface with ease as shown in Figure 8(a).

When the process is limited by mass transport phenomena, the controlling factors are the diffusion rate of the reactant through the boundary layer and the diffusion out through this layer of the gaseous by-products. This usually happens when pressure and temperature are high. As a result, the gas velocity is low as was shown above, the boundary layer is thicker making it more difficult for the reactants to reach the deposition surface. Furthermore, the decomposition reaction occurs more rapidly since the temperature is higher and any molecule that reaches the surface reacts instantly. The diffusion rate through the boundary layer then becomes the rate limiting step as shown in Figure 8(b).

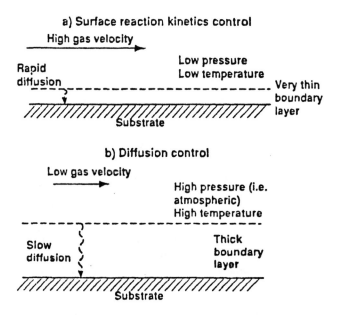

Figure 8. Rate Limiting Steps in a CVD Reaction

As a result, since normally the surface kinetics (or near surface kinetics) is the limiting step at lower temperature and diffusion the rate limiting step at higher temperature, it is possible to switch from one rate limiting step to the other by changing the temperature. This is illustrated in Figure 9 where the Arrhenius plot (logarithm of the deposition rate vs the reciprocal temperature) is shown for several reactions leading to the deposition of silicon, using either SiH_4, SiH_2Cl_2, $SiHCl_3$ or $SiCl_4$ as silicon sources in a hydrogen atmosphere (15). In the A sector (lower right), the deposition is controlled by surface reaction kinetics as the rate limiting step. In the B sector (upper left), the deposition is controlled by the mass transport process and the growth rate is related linearly to the partial pressure of the silicon reactant in the carrier gas. Transition from one rate control regime to the other is not sharp, but involves a transition zone where both are significant. The presence of a maximum in the curves in Area B would indicate the onset of gas-phase precipitation, where the substrate has become starved and the deposition rate decreased.

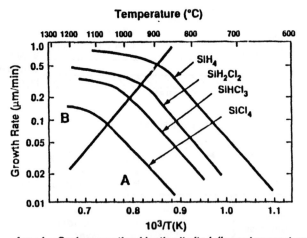

Figure 9. Arrhenius Plot for Silicon Deposition Using Various Precursors

Pressure is similar to temperature as a rate limiting factor since the diffusivity of a gas is inversely related to its pressure. For instance, lowering the pressure from one atmosphere to 1 Torr increases the gas phase transfer of reactants to the deposition surface and the diffusion out of the by-products by more than two orders of magnitude. Clearly, at low pressure, the effect of mass transfer variables is far less critical than at higher pressure.

It can be now seen that, by proper manipulation of the process parameters and reactor geometry, it is possible to control the reaction and the deposition to a great degree. This is illustrated by the following example. In the deposition of tungsten in a tube mentioned in Section Two above, the gas velocity is essentially constant and the boundary layer gradually increases in thickness toward downstream. This means that the thickness of the deposit will decrease as the distance from the tube inlet increases as shown in Figure 10(a). This thickness decrease can be offset and a more constant thickness obtained simply by tilting the susceptor as shown in Figure 10(b). This increases the gas velocity due the flow constriction; the Reynolds number goes up, the boundary layer decreases and the deposition rate is more uniform (14).

3.4 Mathematical Expressions of the Kinetics of CVD

Many workers have successfully expressed the flow dynamics and mass transport processes mathematically and obtained a realistic model that could be used in the predictions of a CVD system and in the design of reactors (16, 17, 18). These models are designed to define the complex entrance effects and convection phenomena that occur in a reactor and solve the complete equations of heat, mass balance and momentum. They can be used to optimize the design parameters of a CVD reactor such as susceptor geometry, tilt angle, flow rates and others. To obtain a complete and thorough analysis, these models should be complemented with experimental observations such as the flow patterns mentioned above and *in situ* diagnostic techniques such as laser Raman spectroscopy (19).

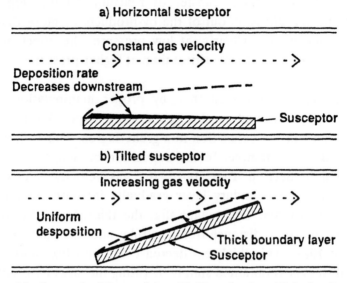

Figure 10. Control of Deposition Uniformity in a Tubular Reactor

4.0 GROWTH MECHANISM AND STRUCTURE OF DEPOSIT

In the previous sections, it was shown how thermodynamic and kinetic considerations govern a CVD reaction. In this section, the nature of the deposit, i.e. its microstructure and how it is controlled by the deposition conditions, will be examined.

4.1 Deposition Mechanism and Epitaxy

The manner in which a film is formed on a surface by CVD is still a matter of controversy and several theories have been advanced to describe the phenomena (2). A thermodynamic theory proposes that a solid nucleus is formed from supersaturated vapor as a result of the difference between the surface free energy and the bulk free energy of the nucleus. Another and newer theory is based on atomistic nucleation and combines chemical bonding of solid surfaces and statistical mechanics (20). These theories are certainly very valuable in themselves but considered outside the scope of this book. There are however three important factors that control the nature and properties of the deposit to some degree which must be reviewed at this time: epitaxy, gas phase precipitation and thermal expansion.

Epitaxy. The nature of the deposit and the rate of nucleation at the very beginning of the deposition are affected, among other factors, by the nature of the substrate. A specific case is that of epitaxy where the structure of the substrate essentially controls the structure of the deposit (2,15,21). Epitaxy can be defined as the growth of a crystalline film on a crystalline substrate, the substrate acting as a seed crystal. When both substrate and deposit are of the same material (for instance silicon on silicon) or when their crystalline structures (lattice parameters) are identical or very close, the phenomena is known as homoepitaxy. When the lattice parameters are different, it is heteroepitaxy. Epitaxial growth cannot occur if these differences are too great. A schematic of epitaxial growth is shown in Figure 11. As an example, it is possible to grow gallium arsenide epitaxially on silicon since their lattice parameters are close. On the other hand, deposition of indium phosphide on silicon is not possible since the lattice mismatch is 8%, which is too high. A solution is to use an intermediate buffer layer of gallium arsenide between the silicon and the indium phosphide. The lattice parameters of common semiconductor materials are shown in Figure 12.

Generally epitaxial films have superior properties and, whenever possible, epitaxial growth should be promoted. The epitaxial CVD of silicon and III-V and II-VI compounds is now a major process in the semiconductor industry and is expected to play an increasingly important part in improving the performance of integrated circuits in the near future as will be shown in Chapters Eight and Nine (22).

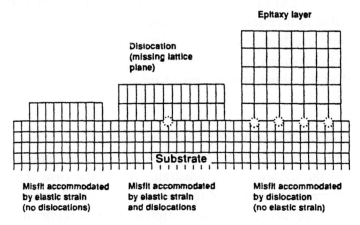

Figure 11. Epitaxy Accommodations of Lattice Mismatch

Gas phase precipitation. As mentioned previously, a CVD reaction may occur in the gas phase instead of at the substrate surface, if the supersaturation of the reactive gases and the temperature are high enough. This is generally detrimental as gas phase precipitated particles in the form of soot become incorporated in the deposit causing non uniformity in the structure, surface roughness and poor adhesion. In some cases, gas phase precipitation is used purposely, for instance in the production of very fine powders (see Chapter Twelve).

Thermal expansion. Large stresses can be generated in a CVD coating during the cooling period from deposition temperature to room temperature, if there is a substantial difference between the coefficients of thermal expansion (CTE) of the deposit and of the substrate. These stresses may cause cracking and spalling of the coating. As an example, it is extremely difficult to coat carbon-carbon (which has almost zero thermal expansion) with ceramics such as silicon carbide, hafnium carbide or hafnia to provide oxidation protection (23). These ceramic materials have higher thermal expansion than carbon-carbon, are brittle and usually microcrack upon cooling. This obviously is an extreme case but it points to the need to match CTE's as closely as possible. If differences are large, it may be necessary to use a buffer coating with an intermediate CTE or with high ductility. Deposition processes which do not require high temperature such as plasma CVD or MOCVD should also be considered (see Chapter Four). Table 1 lists the CTE of typical CVD materials and substrates.

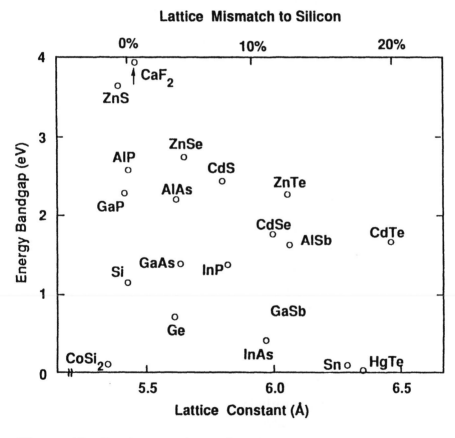

Figure 12. Bandgap and Lattice Constant of Semiconductor Materials

TABLE 1

Coefficient of Thermal Expansion (CTE) of Typical CVD Materials and Substrates

Materials	CTE (ppm/°C) 25-300°C
Metals	
Aluminum	23.5
Gold	14.2
Iridium	6
Molybdenum	5
Niobium	7
Steel (carbon)	12
Steel, stainless (302)	17.3
Tantalum	6.5
Titanium	9
Tungsten	4.5
Non Metallic Elements	
Carbon	5.4
Carbon-carbon	0.5
Silicon	3.8
Ceramics	
Alumina	8.3
Boron carbide	4.5
Boron nitride	7.5
Chromia	8
Hafnia	7

TABLE 1 (cont.)

Coefficient of Thermal Expansion (CTE) of Typical
CVD Materials and Substrates

Materials	CTE (ppm/°C) 25-300°C
Ceramics (cont.)	
Magnesia	13
Molybdenum disilicide	8.25
Silicon carbide	3.9
Silicon nitride	2.45
Silicon dioxide	0.5
Titanium carbide	7.6
Titanium diboride	6.6
Titanium nitride	9.5
Tungsten carbide	4.5
Tungsten disilicide	6.6

Note: Reported values of CTE's often vary widely. The values listed here were culled from several sources and must be considered as representative and verified for design work.

4.2 Structure and Morphology of CVD Materials

The properties of a CVD material are directly related to the nature of its structure which is in turn controlled by the deposition conditions. In this section and the next, the relationship between properties, structure and deposition conditions will be examined.

The structure of a CVD material can be divided into three major types which are shown schematically in Figure 13 (24). Zone (A) is a structure consisting of columnar grains which are capped by a dome-like top, zone (B) consists also of columnar growths but more faceted and regular and zone (C) is made up of fine equiaxed grains. Examples of these structures are shown in Figure 14 (25). This is the CVD equivalent of the structural model for vacuum-evaporated films first introduced by Movchan and Demshishin (26).

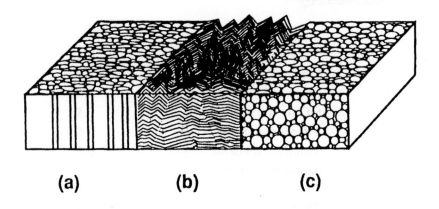

(a)   **(b)**   **(c)**

Figure 13. Schematic of Structures Obtained by CVD:
a) Columnar Grains with Domed Tops
b) Faceted Columnar Grains
c) Equiaxed Fine Grains

As might be expected, the microstructure varies depending on the material being deposited. In general ceramics obtained by CVD such as SiO_2, Al_2O_3, Si_3N_4 and most dielectric materials tend to be amorphous or, at least, have a very small grain microstructure (Type C). Metal deposits tend to be more crystalline with the typical columnar structure of type (A) or (B). Deposit crystal size is also a function of deposition conditions, especially temperature.

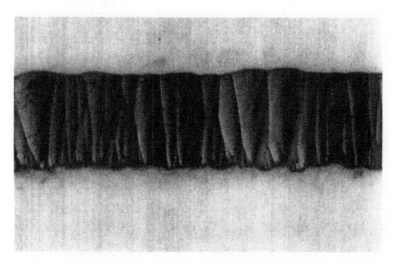

Figure 14. Examples of CVD Structures:
a) Columnar Grains with Domed Tops
Source: Ultramet, Pacoima CA

Figure 14. Examples of CVD Structures:
b) Faceted Columnar Grains
Source: Ultramet, Pacoima CA

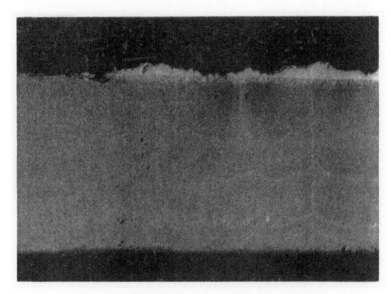

Figure 14. Examples of CVD Structures:
c) Equiaxed Fine Grain
Source: Ultramet, Pacoima CA

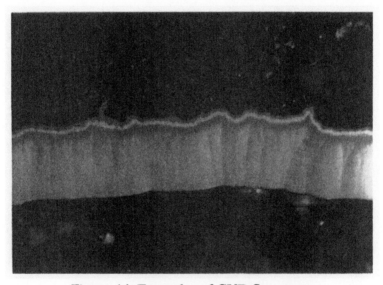

Figure 14. Examples of CVD Structures:
d) Mixed Structures
Source: Ultramet, Pacoima CA

Generally the most desirable structure for load-bearing use is the fine equiaxed (C) which has the highest mechanical properties, hardness and fracture toughness.

More often than not, a deposited structure will include two and sometimes all three types, as shown in Figure 14 (x and y). This usually happens in thick deposits where a uniform structure is more difficult to obtain.

4.3 The Control of CVD Microstructure

It is possible to control the nature of a CVD structure by the proper manipulation of the deposition parameters such as temperature, pressure, supersaturation and the selection of the CVD reaction.

Pressure controls the thickness of the boundary layer and consequently the degree of diffusion as was shown in Section Three above. By operating at low pressure, the diffusion process can be minimized and surface kinetics becomes rate controlling. Under these conditions, deposited structures tend to be fine-grained which is usually a desirable condition. It is the structure shown in Figure 13, Zone C.

Fine-grained structures can also be obtained at low temperature and high supersaturation as well as low pressure. At higher temperatures, deposits tend to be columnar (Figure 13, zones A and B) as a result of uninterrupted grain growth toward the reactant source. The structure is also often dependent on the thickness of the deposit. For instance, grain size will increase as the thickness increases. A columnar-grain structure develops which becomes more pronounced as the film becomes thicker.

Columnar structures are usually undesirable as the deleterious effects of grain growth and columnar formation can be considerable. They may lead to structural, chemical and electrical anisotropy and the rapid diffusion of impurities along the grain boundaries. It is possible to reduce or eliminate columnar deposits and obtain fine grain equiaxed growth by mechanical means such as rubbing or brushing at regular interval to renucleate the deposition surface. This approach has been demonstrated in the deposition of tungsten (13). However it is not generally practical, particularly when substrates of complex geometry are used.

Another approach is the control of grain growth by chemical means (27,28). In the deposition of tungsten for instance, a fine grain equiaxial growth is obtained by depositing alternating layers of tungsten and silicon. Typically a layer of tungsten approximately 1000 Angstroms thick is deposited using the hydrogen reduction of WF_6 at $550°C$ and a pressure of a few Torr as follows:

[1] $WF_6 + 3H_2 \longrightarrow W + 3HF$

A 150Å thick layer of silicon is then deposited by the thermal decomposition of silane at the same temperature and pressure as follows:

[2] $SiH_4 \longrightarrow Si + 2H_2$

WF_6 is then reintroduced in the system and the thin layer of silicon is reduced as follows:

[3] $3Si + 2WF_6 \longrightarrow 2W + 3SiF_4$

After the silicon is completely reduced, reaction (3) stops and reaction (1) is resumed. The temporary deposition of sacrificial silicon interrupts the tungsten grain growth as new nucleation sites are created. Larger grain and columnar growth are essentially eliminated and a fine grained structure is the result. A similar process can be used for the deposition of rhenium and other metals.

REFERENCES

1. Spear, K.E., Thermochemical Modeling of Steady-state CVD Process, *Proc. 9th. Int. Conf. on CVD*, (McD. Robinson et al, Eds.), pp. 81-97, Electrochem. Soc., Pennington NJ 08534 (1984)

2. Kern, W. and Ban, W.S., Chemical Vapor Deposition of Inorganic Thin Films, in *Thin Film Processes,* (J.L. Vossen and W. Kern, Eds.), Academic Press, 257-331, New York (1978)

3. Gokoglu, S.A., Chemical Vapor Deposition Modeling- An Assessment of Current Status, Proc. 11th. Int. Conf. on CVD (K. Spears and G. Cullen, Eds.) pp.1-9, *Electrochem. Soc.*, Pennington, NJ 08534 (1990)

4. Powell, C.F., Oxley, J.H. and Blocher J.M. Jr., *Vapor Deposition*, John Wiley and Sons, New York (1966)

5. Pierson, H.O., The CVD of TiB_2 from Diborane, *Thin Solid Films*, 72:511-516 (1980)

6. Chase, M.W., *JANAF Thermochemical Tables*, Vol.13, Supp. No.1, American Chem. Soc. & Am. Inst. of Physics (1985)

7. Many sources of thermodynamic data are now available including the following:

• Wagman, D.D., *The NBS Tables of Chemical Thermodynamic Properties*, J. Phys. Chem. Ref. Data 11 and supplements (1982)

• Barin, I and Knacke, O., *Thermochemical Properties of Inorganic Substances*, Springer, Berlin (1983)

• Hultgren, R., *Selected Values of the Thermodynamic Properties of the Elements*, Am. Soc. for Metals, Metals Park, OH (1973)

• Mills, K.C., *Thermodynamic Data for Inorganic Sulphides, Selenides and Tellurides*, Butterworth, London (1974)

8. Nolang, B., Equilibrium Computations and their Application to CVD Systems, *Proc. of 5th. European Conf. on CVD* , (J. O. Carlsson and J. Lindstrom, Eds.), pp. 107-115, University of Uppsala, Sweden (1985)

9. Gordon, S. and McBride, B.J., NASA Publication SP273 (1971)

10. Eriksson, G., *Acta Chem. Scandia*, 25:2651-2658 (1971)

11. Besmann, T.M., SOLGASMIX-PV, A Computer Program to Calculate Equilibrium Relationship in Complex Chemical Systems, ORNL/TM-5775, *Oak Ridge National Laboratory,* Oak Ridge, TN (1977)

12. EKVICALC and EKVIBASE, *Svensk Energi Data*, Agersta, S-740 22 Balinge, Sweden

13. Blocher, J.M. Jr., Chemical Vapor Deposition, in *Deposition Technologies for Films and Coatings*, (R. Bunshah, Ed.), pp. 348-351, Noyes Publications, Park Ridge, NJ (1982)

14. Sherman, A., *Chemical Vapor Deposition for Microelectronics,* Noyes Publications, Park Ridge, NJ (1987)

15. Pearce, C.W., Epitaxy, in *VLSI Technology*, (S. M. Sze Ed.), McGraw-Hill Book Co., New York (1983)

16. Jensen, K.F., Modelling of Chemical Vapor Deposition Reactors, *Proc. 9th. Int. Conf. on CVD*, (McD. Robinson et al, Eds.), pp. 3-20, Electrochem. Soc., Pennington NJ 08534 (1984)

17. Rosenberger, F., Flow Dynamics and Modelling of CVD, *Proc. 10th. Int. Conf. on CVD*, (G. W. Cullen, Ed.), pp. 193-203, Electrochem. Soc., Pennington, NJ 09534 (1987)

18. He, Y. and Sahai, Y., Three-dimensional Mathematical Modeling of Transport Processes in CVD Reactors, in *Proc. 10th. Int. Conf. on CVD*, (G. W. Cullen, Ed.), Electrochem. Soc. Pennington, NJ (1987)

19. Breiland, W.G.,Ho, P. and Coltrin, M.E., Laser Diagnostics of Silicon CVD: In Situ Measurements and Comparisons with Model Predictions, *Proc. 10th. Int. Conf. on CVD*, (G. W. Cullen, Ed.), 912-924. Electrochem. Soc. Pennington, NJ 09534 (1987)

20. Gretz, R.D. and Hirth, J.P., Nucleation and Growth Process in CVD, in *CVD of Refractory Metals Alloys and Compounds*, pp. 73-97, Am. Nuclear Soc., Hinsdale, Ill. (1967)

21. Venables, J.A. and Price, G.L., in *Epitaxial Growth*, (J. W. Matthews, Ed.), Pt.B, p.382, Academic Press, New York (1975)

22. Hammond, M.L. and Sandler, N.P., Epitaxial Silicon: A Vital Process for the 1990s, *Proc. 11th Int. Conf. on CVD*, (K. Spear and G. Cullen, Eds.), 277-283, Electrochem. Soc., Pennington, NJ 08534 (1990)

23. Pierson, H.O., Sheek, J.G. and Tuffias, R.H., Overcoating of Carbon-carbon Composites, Final Report, WRDC-TR-89-4045,

Wright-Patterson AFB, OH 45433-6533 (1989)

24. Stinton, D.P., Bessman, T.M. and Lowden, R., Advanced Ceramics by Chemical Vapor Deposition Techniques, *Cer. Bul.*,67-2:350-355 (1988)

25. Photographs by permission of Ultramet, Pacoima CA 91331

26. Movchan, B.A. and Demchishin, A.V., *Fisika Metall.* 28, 653 (1959)

27. Green, M.L., et al, The Formation and Structure of CVD W Films Produced by the Si Reduction of WF_6, *J. Electrochem. Soc.*,134-9:2285-92 (1987)

28. Kamins T.I. et al, *J. Electrochem. Soc.*,133-12:2555-9 (1986)

3

THE CHEMISTRY OF CVD

1.0 CATEGORIES OF CVD REACTIONS

As could be expected, the fundamental element in CVD is the chemical reaction and, before any CVD program is undertaken, it is essential to evaluate all the reactions and select the most appropriate. This is done by a thermodynamic analysis as described in Chapter Two and, if necessary, by an experimental program.

The CVD reactions can be classified into several major categories which are outlined below. These reactions and others will be described in greater detail in Chapters Five, Six and Seven where CVD materials are reviewed individually.

1.1 Thermal Decomposition (or Pyrolysis) Reactions

In thermal decomposition reactions, a molecule is broken apart into its elements and/or a more elementary molecule. Such reactions are the simplest since only one precursor gas is used. Typical examples are as follows:

Hydrocarbon decomposition:

[1] $CH_4(g) \longrightarrow C(s) + 2H_2(g)$

This reaction is used extensively in the production of carbon, graphite and diamond.

Hydride decomposition (see Section 6):

[2] $SiH_4(g) \longrightarrow Si(s) + 2H_2(g)$

[3] $B_2H_6(g) \longrightarrow 2B(s) + 3H_2(g)$

Carbonyl decomposition (see Section 4):

[4] $Ni(CO)_4(g) \longrightarrow Ni(s) + 4CO(g)$

This also includes the decomposition of the more complex carbonyls such as the carbonyl hydrides and nytrosylcarbonyls.

Metallo-organic decomposition (see Section 5):

[5] $CH_3SiCl_3(g) \longrightarrow SiC(s) + 3HCl(g)$

Such reactions are becoming very important particularly in the semiconductor industry.

Halide decomposition (see Section 3):

[6] $WF_6(g) \longrightarrow W(s) + 3F_6(g)$

[7] $TiI_4(g) \longrightarrow Ti(s) + 2I_2(g)$

1.2 Hydrogen Reduction

Reduction is a chemical reaction in which an element gains an electron, in other words when the oxidation state is lowered (1). Reduction reactions are very important CVD reactions. This is particularly true of the hydrogen reduction of the halides, and the following examples are used widely:

[8] $WF_6(g) + 3H_2(g) \longrightarrow W(s) + 6HF(g)$

[9] $SiCl_4(g) + 2H_2(g) \longrightarrow Si(s) + 4HCl(g)$

These two reactions as shown are actually simplified and several intermediate reactions generally occur with the formation of lower halides.

Hydrogen reduction has a major advantage in that the reaction generally takes place at lower temperature than the equivalent decomposition reaction. It is used extensively in the deposition of transition metals from their halides, particularly the metals of the groups Va, VIa and VIIa, such as vanadium, tantalum, molybdenum, tungsten and rhenium. The halide reduction of the group IVa metals (titanium, zirconium and hafnium) however is more difficult because their halides are more stable.

The hydrogen reduction of the halides of non metallic elements such as boron and silicon (Reaction 9) is used widely in the deposition of semiconductors and the production of fibers.

Hydrogen is used in a supplementary role in reactions where reduction is not the primary function, for instance where it is necessary to prevent the formation of oxides or carbides (for the weak members of those classes such as FeO) and generally improve the characteristics and properties of the deposited material.

1.3 Coreduction

The deposition of a binary compound can be achieved by a coreduction reaction. In this manner, ceramic materials such as oxides, carbides, nitrides, borides and silicides can be produced readily, providing of course that the reaction is favorable. In these cases, the compound will be obtained more readily than the metal. A common example is the deposition of titanium diboride:

[10] $TiCl_4(g) + 2BCl_3(g) + 5H_2(g) \longrightarrow TiB_2(s) + 10HCl(g)$

1.4 Metal Reduction of the Halides

Although hydrogen is the most common reductant, there are other elements which are even more powerful reductants such as zinc, cadmium, magnesium, sodium and potassium, as shown in Table 1.

In this table, the free energy of formation, ΔG_f^0, of the chloride of these metals is listed for four different temperatures. As can be seen, the values are more negative than that of hydrogen chloride. These metals can be used to reduce the halides of titanium, zirconium or hafnium, whereas hydrogen, as mentioned above, cannot do so readily. In order to be useful in CVD, the by-product chloride must be volatile at the deposition temperature. This may rule out the use of sodium or potassium which evaporate above 1400°C.

Of these reductant metals, the most commonly used is zinc. The reason is that the zinc halides are more volatile than the parent metal and the chances of codeposition of the halides are minimized. Either chloride or iodide are used, although the iodide, being the most volatile, is usually preferred. The volatility of the halides of these elements decreases as one goes from the iodides to the chlorides to the fluorides. The reaction is as follows:

[11] $TiI_4(g) + 2Zn(s) \longrightarrow Ti(s) + 2ZnI_2(g)$

TABLE 1

Free Energy of Formation of Reductant Elements (2)

Reductant	Melt Point °C	Boil Point °C	Chloride	Standard Free Energy of Formation (KJoule per mole) at the given temperatures(°C)			
				425	725	1025	1325
H_2	-	-	HCl	-98	-100	-102	-104
Cd	320	765	$CdCl_2$	-264	-241	-192	-164
Zn	419	906	$ZnCl_2$	-307	-264	-262	-254
Mg	650	1107	$MgCl_2$	-528	-482	-435	-368
Na	97	892	NaCl	-223	-238	-241	-229
K	631	760	KCl	-369	-341	-295	-

Another reductant, magnesium, is used in the industrial production of titanium metal as follows:

[12] $TiCl_4(g) + 2Mg(s) \longrightarrow Ti(s) + 2MgCl_2(g)$

The alkali metals, sodium and potassium, shown in Table 1 are not generally used as reductants because their reductive power is so high that it tends to cause premature and detrimental gas phase precipitation and, as mentioned above, the high temperature required to volatilize the by-products is a major negative factor.

1.5 Oxidation and Hydrolysis Reactions

Oxidation and hydrolysis are two important groups of reactions which are used in the formation of oxides. Common sources of oxygen are the element itself and CO_2. Recently, ozone (O_3) has been used for the deposition of SiO_2 (3). It is a very powerful oxidizing agent which is generally generated in situ, by a corona discharge in oxygen.

Typical oxidation and hydrolysis reactions are:

[13] $SiH_4(g) + O_2(g) \longrightarrow SiO_2(s) + 2H_2(g)$

[14] $SiCl_4(g) + 2CO_2(g) + 2H_2(g) \longrightarrow SiO_2(s) + 4HCl(g) + 2CO(g)$

[15] $2AlCl_3(g) + 3H_2O(g) \longrightarrow Al_2O_3(s) + 6HCl(g)$

1.6 Reactions to Form Carbides and Nitrides

The deposition of carbides (or carbidization) is usually obtained by reacting a halide with a hydrocarbon such as methane as follows:

[16] $TiCl_4(g) + CH_4(g) \longrightarrow TiC(s) + 4HCl(g)$

The deposition of nitrides (nitridation or ammonolysis) is generally based on ammonia which is preferred to nitrogen. Ammonia has a positive free energy of formation; thus its equilibrium products are essentially hydrogen and nitrogen which become the reactants for the CVD reaction. The observed advantage of ammonia as a reactant lies in the kinetics of the nascent H and N from the ammonia decomposition. Thermo-dynamically, the reaction of $2NH_3$ and of N_2 and H_2 are equivalent.

An example of the use of ammonia is the deposition of silicon nitride, a reaction that is widely used in the semiconductor industry:

[17] $3SiH_4(g) + 4NH_3(g) \longrightarrow Si_3N_4(s) + 12H_2(g)$

2.0 CVD PRECURSORS

In the previous section, the various general categories of CVD chemical reactions were reviewed and, as can be readily seen, the choice of the proper reactants (the precursors) is very important. These precursors fall into several general major groups which are the halides, carbonyls, metallo-organics and hydrides. A general review of the nature and properties of these precursors is given in this section which should help in making the proper precursor selection and gaining a better understanding of the CVD reactions.

The choice of a precursor is governed by certain general characteristics which can be summarized as follows:

- stability at room temperature

- ability to react cleanly in the reaction zone

- sufficient volatility at low temperature so that it can be easily transported to the reaction zone without condensing in the lines

• capability of being produced in a very high degree of purity

• ability to react without producing side reactions or parasitic reactions

3.0 HALIDES

The halides are binary compounds of a halogen (elements of group VIIb of the periodic table) and a more electropositive element such as a metal.

3.1 Halogens

The halogens include fluorine, chlorine, bromine and iodine and all have been used in CVD reactions. They are very reactive elements and exist as diatomic molecules, i.e. F_2, Cl_2 etc.. Their relevant properties are listed in Table 2.

Of all the elements, fluorine is the most chemically reactive. It combines directly with other elements. Chlorine is slightly less reactive . Both are gases at room temperature which is an important advantage in CVD systems. Because of their reactivity, they form halides readily but also attack most materials which makes them difficult to handle and requires equipment designed with inert materials such as Monel or Teflon. Halogens are also toxic, fluorine more so than chlorine by an order of magnitude.

Bromine is a dark-red liquid with high specific gravity. Iodine is a black solid which sublimes at atmospheric pressure producing a violet vapor. They are used in CVD but to a lesser degree than either fluorine or chlorine.

TABLE 2

Relevant Properties of the Halogens (4)

Element	Electron Affinity KJ g-atom	Boiling Point °C	Melting Point °C
Fluorine	339	-118	-233
Chlorine	355	-34.6	-103
Bromine	331	58.7	-7.2
Iodine	302	184.3	113.5

3.2 Halide Formation or Halogenation

Halides can be formed by the direct interaction of the halogen as follows:

[1] $W(s) + 3Cl_2(g) \longrightarrow WCl_6(g)$

or by the reaction with another halide:

[2] $W(s) + 6HCl(g) \longrightarrow WCl_6(g) + 3H_2(g)$

Reaction 2 is often preferred because of the hazard of unconsumed chlorine from Reaction 1 reacting explosively with the hydrogen commonly used as a reducing agent.

3.3 Halide Properties

Properties of some halides useful in CVD are in Table 3.

As can be seen, some of these halides are gaseous or liquid at room temperature and are easily transported in the reaction chamber. But the solid halides must be heated to produce sufficient vapor. This sometimes presents a problem which can be bypassed by generating the halide *in situ*, using the process described in Chapter Four, Section 3. Most halides are available commercially.

TABLE 3

Properties of Selected Halides (4)

Halides	Melting Point $^\circ$C	Boiling Point $^\circ$C
$AlBr_3$	97.5	263
$AlCl_3$	190	182.7 (s)
BCl_3	-107.3	12.5
BF_3	-126.7	-99.9
CCl_4	-23	76.8
CF_4	-184	-128
$CrCl_2$	824	1300 (s)
$HfCl_4$	319	319 (s)
HfI_4		400 (s)
$MoCl_5$	194	268
MoF_6	17.5	35
$NbCl_5$	204.7	254
ReF_6	18.8	47.6
$SiCl_4$	-70	57.6

TABLE 3 (cont.)

Properties of Selected Halides (4)

Halides	Melting Point °C	Boiling Point °C
$TaBr_5$	265	348.8
$TaCl_5$	216	242
$TiCl_4$	-25	136
VCl_4	-28	148.5
WCl_5	248	275.6
WF_6	2.5	17.5
$ZrBr_4$	450	357 (s)
$ZrCl_4$	437	331 (s)

Note: (s) indicates that the compound sublimes at atmospheric pressure before melting.

4.0 METAL CARBONYLS

The metal carbonyls are a large and very important group of compounds which are used widely in the chemical industry, particularly in the preparation of heterogeneous catalysts and as precursors in CVD and metallo-organic CVD (MOCVD).

4.1 Characteristics of the Carbonyls

The carbonyls are relatively simple compounds since they consist of only two basic components: a) the carbonyl group which is a functional group where a carbon atom is doubly bonded to an oxygen atom (-CO-), and b) a d-group transition metal. The atomic structure of these metals is such that the d shell is only partly filled. The first transition series (3d) comprises Sc, Ti, V, Cr, Mn, Fe, Co and Ni; the second (4d), Y, Zr, Nb, Mo, Tc, Ru, Rh, Pd and Ag; the third (5d), Hf, Ta, W, Re, Os, Ir, Pt and Au. Carbonyl derivatives of at least one type are found for all these metals. Although only a few are presently used in CVD, many are being investigated as they constitute an interesting and potentially valuable source of precursor materials.

The simplest transition metal carbonyls are mononuclear of the type $M(CO)_x$, in other words those with only one metal atom. They are hydrophobic but soluble to some extent in non-polar liquids such as n-butane or propane. The dinuclear carbonyls are more complex but have the same general characteristics as the mononuclear carbonyls. The carbonyls, which are or could be used in CVD, are listed in Table 4 with some of their properties.

TABLE 4

Selected Properties of Metal Carbonyls (1,5)

Compound	MP °C	BP °C	SpGr	Color and Form	Structure	Comments
MONONUCLEAR						
$V(CO)_6$	65	dec	-	Black crystal	Octahedral	Yellow-Orange in solution, volatile, very unstable

TABLE 4 (cont.)

Selected Properties of Metal Carbonyls (1,5)

Compound	MP °C	BP °C	SpGr	Color and Form	Structure	Comments
$Cr(CO)_6$	164	dec 180	1.77	White crystal	Octahedral	Volatile
$Fe(CO)_5$	-20	103	1.42	Yellow liquid		Poisonous
$Ni(CO)_4$	-25	43	1.32	Colorless liquid		Very toxic
$Mo(CO)_6$	150	dec 180		White crystal	Octahedral	Volatile
$Ru(CO)_5$	-22	-		Colorless liquid		Very volatile
$W(CO)_6$	169	dec	2.65	White crystal	Octahedral	Volatile
$Os(CO)_5$	-15	-		Colorless liquid		Very volatile
DINUCLEAR						
$Mn_2(CO)_{12}$	152	-	-	Crystal		
$Fe_2(CO)_9$	80 dec	-	2.08	Yellow crystal	Hexagonal	Volatile

TABLE 4 (cont.)

Selected Properties of Metal Carbonyls (1,5)

Compound	MP °C	BP °C	SpGr	Color and Form	Structure	Comments
$Co_2(CO)_8$	51	52 dec	1.73	Orange crystal		CO loss
$Re_2(CO)_{10}$	170	250 dec		Colorless crystal	Cubic	Volatile
$Ir_2(CO)_8$	160	-		Yellow crystal		Sublimes in in CO_2 at 160°C

More complex yet are the polynuclear or cluster carbonyls, some of which are used as CVD precursors such as: $Fe_3(CO)_{12}$, $Ru_3(CO)_{12}$, $Os_3(CO)_{12}$, $Co_4(CO)_{12}$, $Rh_4(CO)_{12}$ and $Ir_4(CO)_{12}$. Some of these compounds are used in catalyst preparation, either as pure metal or metal with attached CO group.

4.2 Carbonyl Preparation

Many metal carbonyls are available commercially. However, in some cases, the CVD investigator may find it more expedient (and sometimes cheaper) to produce them in-house. This is particularly true of the only two carbonyls that can be obtained by the direct reaction of the metal with CO (and consequently easy to synthesize), i.e. nickel carbonyl, $Ni(CO)_4$, and iron carbonyl

$Fe(CO)_5$. Table 5, below, summarizes the various synthesis processes.

TABLE 5

Metal Carbonyl Preparation

Carbonyl	Process
$Ni(CO)_4$	Direct reaction with CO at one atm. and at 80°C
$Fe(CO)_5$	Direct reaction with CO at one atm. and at 150-300°C
$Co_2(CO)_8$	Reaction of the carbonate with CO at 250-300 atm. and at 120-200°C
$V(CO)_6$ $Cr(CO)_6$ $W(CO)_6$ $Re_2(CO)_{10}$	Reaction of the chloride with CO with a reducing agent at 200-300 atm. and at 300°C
$Os(CO)_5$ $Re_2(CO)_{10}$	Reaction of the oxide with CO at 300°C
$Ru_3(CO)_{12}$	Reaction of acetylacetonate with CO at 150°C and 200 atm.

4.3 Metal Carbonyl Complexes

Metal carbonyls form several complexes but those of major interest in MOCVD are the carbonyl halides and the carbonyl nitric oxide complexes.

The carbonyl halides have the general formula $M_x(CO)_yX_z$ with X being fluorine, chlorine, bromine or iodine. With the notable exception of nickel, they exist for most of the metals that form carbonyls and for all those listed in Table 4.

A particularly interesting case is that of the platinum metals group which comprises ruthenium (Ru), osmium (Os), rhodium (Rh), iridium (Ir), palladium (Pd) and platinum (Pt). The carbonyl halides of these metals are usually the most practical precursors for metal deposition because of their high volatility at low temperature. Indeed two of them, paladium and platinum, do not form carbonyls but only carbonyl halides. This is also true of gold.

The hydrogen reduction of the metal halides, described in Section 1.2 is generally the favored reaction for metal deposition but is not suitable for the platinum group metals since the volatilization and decomposition temperatures of their halides are too close to provide efficient vapor transport (2). For that reason, the decomposition of the carbonyl halide is preferred. The exception is palladium which is much more readily deposited from the halide reduction than from the carbonyl halide decomposition.

The carbonyl halides are generally produced by the direct reaction of the metal halide and CO, usually at high pressure and at the temperature range of 140-290°C. A typical reaction is (6):

[1] $2PtCl_2(s) + 2CO(g) \longrightarrow 2Pt(CO)Cl_2(s)$

Some examples of carbonyl halides are shown in Table 6.

TABLE 6

Metal Carbonyl Halides (1)

Compound	MP °C	BP °C	Sp.Gr.	Color & Form	Comments
$Mn(CO)_5Cl$	-	-	-	Yellow	Loses CO at 120°C in organic solvent
$Re(CO)_4Cl_2$	-	250 dec.	-	White crystal	
$Ru(CO)_2I_2$	-	-	-	Orange powder	Suitable for CVD
$Os(CO)_3Cl_2$	270	280 dec.	-	Colorless prism	Suitable for CVD
$RhCl_2RhO(CO)_3$	125 subl.	-	-	Ruby red needle crystal	Suitable for CVD
$Ir(CO)_2Cl_2$	140 dec.	-	-	Colorless needle crystal	Suitable for CVD
$Pt(CO)Cl_2$	195	300 dec.	4.23	Yellow needle crystal	Sublimes in CO_2 at 240°C Suitable for CVD
$Pt(CO)_2Cl_2$	142	210	3.48	Light yellow needle crystal	Sublimes in CO_2 at 210°C

Another group of metal carbonyl complexes, worthy of investigation as CVD precursors, consists of the carbonyl nitric oxides. In these complexes, one (or more) CO group is replaced by NO.

An example of these is cobalt nitrosyl tricarbonyl, $CoNO(CO)_3$, which is a preferred precursor for the CVD of cobalt. It is a liquid with a boiling point of 78.6°C which decomposes at 66°C. It is prepared by passing NO through an aqueous solution of cobalt nitrate and potassium cyanide and potassium hydroxide (7).

5.0 METALLO-ORGANIC PRECURSORS

The metallo-organics are relative newcomers as commercially attractive CVD precursors since their first reported use in CVD was in the sixties for the deposition of indium phosphide and indium antimonide. There are now a number of these of prime importance in a branch of CVD known as metallo-organic CVD (MOCVD) which is used extensively in semiconductor and opto-electronic applications. In this section, their chemistry and properties are reviewed (8).

Metallo-organics are compounds in which the atom of an element is bound to one or more carbon atoms of an organic hydrocarbon group. Most of the elements used in MOCVD are from the groups IIa, IIb, IIIb, IVb, Vb and VIb, which are non-transitional. The metallo-organics thus complement the halides and carbonyls which are the precursors for the deposition of transition metals and their compounds.

The term metallo-organic is used somewhat loosely in CVD parlance, since it includes compounds of elements such as silicon, phosphorus, arsenic, selenium and tellurium that are definitely not metallic. To conform to what appears to be a well-established trad-ition, such non-metal compounds will be included here as metallo-organics.

In the following sections, some of the metallo-organic compounds used in CVD are reviewed. Others are described in the review of CVD materials of Chapters Five, Six and Seven. These metallo-organics form only a small proportion of the total number of compounds available and there are many that could be profitably investigated. The main advantage of metallo-organic reactions is a lower deposition temperature.

5.1 Alkyls

The major MOCVD group is that of the alkyls. These are formed by reacting an aliphatic hydrocarbon or an alkyl halide with a metal. These hydrocarbons are composed of chains of carbon atoms as shown below (9):

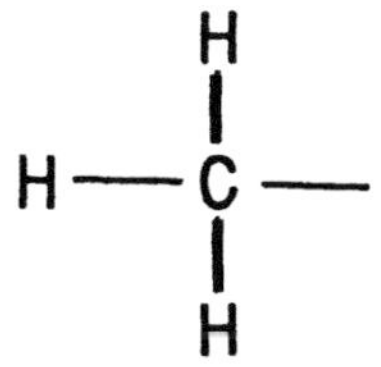

[Methyl]

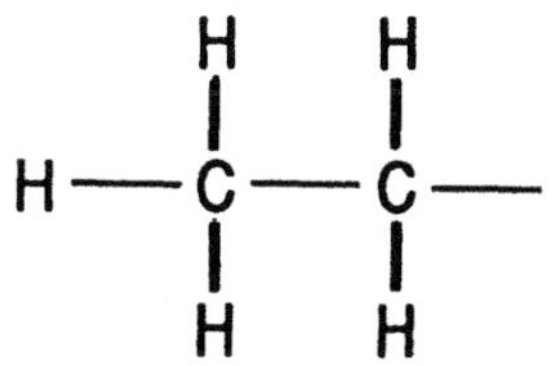

[Ethyl]

$$H - \overset{\overset{\displaystyle H}{|}}{\underset{\underset{\displaystyle H}{|}}{C}} - \overset{\overset{\displaystyle H}{|}}{\underset{\underset{\displaystyle}{|}}{C}} - \overset{\overset{\displaystyle H}{|}}{\underset{\underset{\displaystyle H}{|}}{C}} - H$$

[Isopropyl]

Common metallo-organic alkyls and their relevant properties are listed in Table 7. These alkyls are non-polar volatile liquids. The methyl metallo-organics start to decompose at 200°C and the ethyl metallo-organics at approximately 110°C.

Other common MOCVD compounds are produced from alicyclic hydrocarbons where the carbon chain forms a ring, such as in cyclopentane shown below:

$$CH_2 - CH_2$$

CH₂ ... CH₂ ... CH₂ (ring structure)

Other MOCVD compounds are the aryls which are formed from aromatic hydrocarbons, that is compounds that have six-member rings with three carbon carbon double bonds such as phenyl shown below:

TABLE 7

Metallo-Organic Compounds (1)

Compound	Formula	Type	MP °C	BP °C	Vapor Pressure mm
Trimethyl aluminum	$(CH_3)_3Al$	Alky	15	126	8.4/20°C
Triethyl aluminum	$(C_2H_5)_3Al$	Alky	-58	194	
Triisobutyl aluminum	$(C_4H_9)_3Al$	Alkyl	4	130	
Diisobutyl aluminum hydride	$(C_4H_9)_2AlH$	Alkyl	-70	118	
Trimethyl arsenic	$(CH_3)_3As$	Alkyl			238/20°C
Diethyl arsine	$(C_2H_5)_2AsH_2$	Alkyl			0.8/18°C
Diethyl beryllium	$(C_2H_5)_2B$	Alkyl	12	194	
Diphenyl beryllium	$(C_6H_5)_2Be$	Aryl			
Dimethyl cadmium	$(CH_3)_2Cd$	Alkyl	4	105	28/20°C

TABLE 7 (cont.)

Metallo-Organic Compounds (1)

Compound	Formula	Type	MP °C	BP °C	Vapor Pressure mm
Trimethyl gallium	$(CH_3)_3Ga$	Alkyl	-15	5	64/0°C
Triethyl gallium	$(C_2H_5)_3Ga$	Alkyl	-82	143	18/48°C
Dimethyl mercury	$(CH_3)_2Hg$	Alkyl		96	
Cyclopentadienyl mercury	$(C_5H_5)_2Hg$	Cyclic			
Trimethyl indium	$(CH_3)_3In$	Alkyl	88	134	1.7/20°C
Triethyl indium	$(C_2H_5)_3In$	Alkyl	-32	184	3/53°C
Diethyl magnesium	$(C_2H_5)_2Mg$	Alkyl			
Cyclopentadienyl magnesium	$(C_5H_5)_2Mg$	Cyclic	176		
Ter-butyl phosphine	$(C_4H_9)PH_2$	Alkyl			285/23°C
Triethyl phosphorus	$(C_2H_5)_3P$	Alkyl			10.8/20°C
Tetramethyl lead	$(CH_3)_4Pb$	Alkyl			
Tetraethyl lead	$(C_2H_5)_4Pb$	Alkyl			
Diethyl sulfur	$(C_2H_5)_2S$	Alkyl			
Trimethyl antimony	$(CH_3)_3Sb$	Alkyl			
Trimethyl tin	$(CH_3)_3Sn$	Alkyl			
Cyclopentadienyl tin	$(C_5H_5)_2Sn$	Cyclic			
Diethyl telluride	$(C_2H_5)_2Te$	Alkyl			7/20°C
Dimethyl zinc	$(CH_3)_2Zn$	Alkyl	-42	46	124/0°C
Diethyl zinc	$(C_2H_5)_2Z$	Alkyl	-28	118	6.4/20°C

Note: These compounds are very reactive, especially the lower alkyls. Most of them are volatile, pyrophyric and in some cases react explosively on contact with water. Some are toxic. Manufacturer's recommendations must be carefully studied before use.

Most metallo-organic compounds are monomers with some important exception such as trimethyl aluminum which is a dimer. Their vapor pressures are usually directly related to the molecular weight with the lower molecular weight compounds having the higher volatility.

To be useful as CVD precursors, the metallo-organic compounds should be stable at room temperature so that their storage and transfer are not a problem. They should also decompose readily at low temperature, i.e. below 500°C.

The compounds listed in Table 7 meet these conditions with the exception of the alkyls of arsenic and phosphorus which decompose at higher temperature. For that reason, the hydrides of arsenic and phosphorus are often preferred as CVD precursors (see Section 6 below). These hydrides however are extremely toxic and environmental considerations may restrict their use.

5.2 Acetylacetonates

Metal acetylacetonates (M-ac.ac), also known as pentane-dionates, are produced by reacting metals and acetylacetone. They have the following chemical structure:

$$M \diagdown \begin{matrix} O - C \diagup CH_3 \\ \quad \diagdown CH_2 \\ O - C \diagup \\ \quad \diagdown CH_3 \end{matrix}$$

The acetylacetonates are stable in air and readily soluble in organic solvents. From this standpoint, they have the advantage over the alkoxides, which, with the exception of the iron alkoxides, are not as easily soluble. They can be readily synthesized in the laboratory and many are available commercially and are used extensively as catalysts. They are also used in CVD in the deposition of metals such as iridium, scandium and rhenium and compounds such as the yttrium barium copper oxide complexes used as superconductors (10,11). Commercially available acetyl-acetonates are shown in Table 8.

TABLE 8

Commercially Available Metal Acetyl Acetonates

Metal	Formula	Form	Melting Point °C
Barium	$Ba(C_5H_7O_2)_2$	Crystal	>320
Beryllium	$Be(C_5H_7O_2)_2$	White Crystal	108

TABLE 8 (cont.)

Commercially Available Metal Acetyl Acetonates

Metal	Formula	Form	Melting Point °C	
Calcium	$Ca(C_5H_7O_2)_2$	White crystal	175	dec.
Cerium	$Ce(C_5H_7O_2)_3$	Yellow crystal	131	
Chromium	$Cr(C_5H_7O2)_3$	Violet crystal	214	
Cobalt	$Co(C_5H_7O_2)_3$	Green powder	240	
Copper	$Cu(C_5H_7O_2)_2$	Blue crystal	230	(s)
Dysprosium	$Dy(C_5H_7O_2)_3$	Powder		
Erbium	$Er(C_5H_7O_2)_3$	Powder		
Gadolinium	$Gd(C_5H_7O_2)_3$	Powder	143	
Indium	$In(C_5H_7O_2)_3$	Cream powder	186	
Iridium	$Ir(C_5H_7O_2)_3$	Orange crystal		
Iron	$Fe(C_5H_7O_2)_3$	Orange powder	179	
Lanthanum	$La(C_5H_7O_2)_3$	Powder		
Lead	$Pb(C_5H_7O_2)_2$	Crystal		
Lithium	$LiC_5H_7O_2$	Crystal	ca 250	
Lutetium	$Lu(C_5H_7O_2)_3$	Powder		
Magnesium	$Mg(C_5H_7O_2)_2$	Powder	dec.	
Manganese	$Mn(C_5H_7O_2)_2$	Buff crystal	180	
Neodymium	$Nd(C_5H_7O_2)_3$	Pink powder	150	
Nickel	$Ni(C_5H_7O_2)_2$	Powder	238	
Palladium	$Pd(C_5H_7O_2)_2$	Orange needle	210	
Platinum	$Pt(C_5H_7O_2)_2$	Yellow needle		

TABLE 8 (cont.)

Commercially Available Metal Acetyl Acetonates

Metal	Formula	Form	Melting Point °C	
Praseodymium	$Pr(C_5H_7O_2)_3$	Powder		
Rhodium	$Rh(C_5H_7O_3)_3$	Yellow crystal		
Samarium	$Sm(C_5H_7O_2)_3$	Powder	146	
Scandium	$Sc(C_5H_7O_2)_3$	Powder	187	
Silver	$AgC_5H_7O_2$	Crystal		
Strontium	$Sr(C_5H_7O_2)_2$	Crystal	220	
Terbium	$Tb(C_5H_7O_2)_3$	Powder		
Thorium	$Th(C_5H_7O_2)_4$	Crystal	171	(s)
Thulium	$Tm(C_5H_7O_2)_3$	Powder		
Vanadium	$V(C_5H_7O_2)_3$	Powder		
Ytterbium	$Yb(C_5H_7O_2)_3$	Powder		
Yttrium	$Y(C_5H_7O_2)_3$	Powder		
Zinc	$Zn(C_5H_7O_2)_2$	Needle	138	(s)
Zirconium	$Zr(C_5H_7O_2)_4$	White crystal	172	

Note: (s) indicates that the compound sublimes at atmospheric pressure before melting.

Also increasingly common as CVD precursors are many halogen-acetylacetonate complexes such as trifluoro-acetylacetonate thorium, $Th(C_5H_4F_3O_2)_4$, used in the deposition of thoriated tungsten for thermionic emitters (12), the trifluoro-acetylacetonates of hafnium and zirconium and the hexafluoro-acetylacetonates of calcium, copper, magnesium, palladium, strontium and yttrium.

6.0 HYDRIDES

Hydrides are an important group of precursors that are used to deposit single elements such as boron or silicon. They are also used in conjunction with metallo-organics to form III-V and II-VI semiconductor compounds as shown in the following examples:

[1] $(CH_3)_3Ga(g) + AsH_3(g) \longrightarrow GaAs(s) + 3CH_4(g)$

[2] $x(CH_3)_3Al(g) + (1-x)(CH_3)_3Ga(g) + AsH_3(g) \longrightarrow$

$$Al_xGa_{1-x}As(s) + 3CH_4(g)$$

Many elements form hydrides but only few hydrides are presently used as CVD precursors. These are the hydrides of elements of groups IIIb, IVb, Vb and VIb. All these hydrides are covalent. They are listed in Table 9.

TABLE 9

Hydrides Used in CVD (Partial List) (1,4)

Element	Name	Formula	Boiling Point °C
As	Arsine	AsH_3	-55
B	Diborane	B_2H_6	-92

TABLE 9 (cont.)

Hydrides Used in CVD (Partial List) (1,4)

Element	Name	Formula	Boiling Point °C
Ge	Germane	GeH_4	-88
N	Ammonia	NH_3	-33
P	Phosphine	PH_3	-87
S	Hydrogen sulfide	H_2S	-60
Sb	Stibine	SbH_3	-17
Se	Hydrogen selenide	H_2Se	3
Si	Silane	SiH_4	-111
Te	Hydrogen telluride	H_2Te	-2

Note: Most of these compounds are extremely toxic. Manufacturer's recommendations must be carefully followed.

REFERENCES

1. Cotton, F.A. and Wilkinson, G., *Advanced Inorganic Chemistry,* Interscience Publishers, New York (1972)

2. Powell, C.F., Oxley, J.H. and Blocher, J.M. Jr., *Vapor Deposition,* John Wiley and Sons, New York (1966)

3. Nishimoto, Y., Tokumasu, N., Fujino, K. and Maeda, K., New Low Temperature APCVD Methods Using Polysiloxane and Ozone, *Proc. 11th. Int. Conf. on CVD,* (K. Spear and G. Cullen, Eds.), pp. 410-417, Electrochem. Soc., Pennington NJ 08534 (1990)

4. *Handbook of Chemistry and Physics,* 65th. Ed., CRC Press, Boca Raton, FL (1985)

5. Fillman, L.M. and Tang, S. C., Thermal Decomposition of Metal Carbonyls, *Thermochimica Acta,* 75:71-84, (1984)

6. Lutton, J.M. and Parry, R.W., The Reaction between Platinum Chlorides and Carbon Monoxide, *ACS Journal,* 76:4271-4274 (1954)

7. Harrison, B. and Tompkins, E.H., *Inorg. Chem.* 1:951 (1962)

8. Zilko, J.L., Metallo-Organic CVD Technology and Equipment, in *Handbook of Thin-Film Deposition Processes and Techniques,* (K. K. Shuegraf, Ed.), Noyes Publications (1988)

9. Cram, D.J. and Hammond, G.S., *Organic Chemistry,* McGraw Hill Book Co., New York (1964)

10. Harding, J.T., Kazarof, J.M. and Appel, M.A., Iridium- coated Rhenium Thrusters by CVD, *NASA Technical Memo. 101309,* NASA Lewis Res. Cent., Cleveland, OH (1988)

11. Barron, A.R., Group IIa Metal Organics as MOCVD- Precursors for High T_c Superconductors, *The Strem Chemiker,* XIII-1, Strem Chemicals, Newburyport MA 01950 (1990)

12. Gärtner, G., Jamiel, P. and Lydtin, H., Plasma-activated

Metalorganic CVD of IIIb Oxides/Tungsten Layer Structures, *Proc. of 11th Int. Conf. on CVD*, (K. Spear and G. Cullen, Eds.) pp. 589-595, Electrochem. Soc., Pennington, NJ 08534 (1990)

4

CVD PROCESSES AND EQUIPMENT

1.0 INTRODUCTION

The factors controlling a CVD process are: a) thermodynamic, mass transport and kinetic considerations, which were reviewed in Chapter Two, b) the chemistry of the reaction, which was reviewed in Chapter Three and c) the processing parameters of temperature, pressure and chemical activity which are reviewed in this chapter.

The various CVD processes comprise what is generally known as thermal CVD, which is the original process, and plasma, laser and photo CVD, which are the more recent variations. The difference between these processes is the method of applying the energy required for the CVD reaction to take place as will be discussed later.

2.0 OPEN AND CLOSED REACTOR SYSTEMS

A CVD reaction can occur in one of two basic systems: the closed reactor or the open reactor (also known as closed or open tube). The closed reactor system, also known as chemical transport, was the first type to be used for the purification of metals as mentioned in Chapter One, Section 2. As the name implies, the chemicals are loaded in a container which is then closed. A temperature differential is then applied which provides the driving force for the reaction.

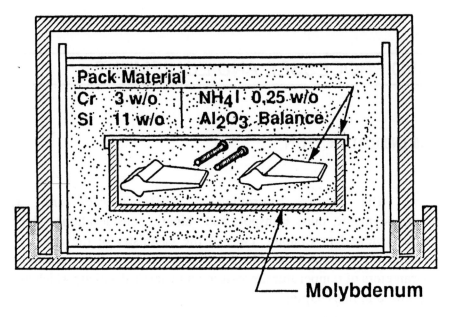

Pack Material
Cr   3 w/o   NH$_4$I 0,25 w/o
Si 11 w/o   Al$_2$O$_3$ Balance

Molybdenum

Figure 1. Schematic of Pack Cementation Chromizing Apparatus

An example of closed tube reaction is the purification of chromium by the iodide process, as follows:

[1]   CrI$_2$ $\longrightarrow$ Cr + 2I (at higher temperature, i.e. $\geq$ 950°C)

[2]   Cr + I$_2$ $\longrightarrow$ CrI$_2$ (at lower temperature, i.e. 750°C)

Note: At higher temperature (reaction 1), iodine is almost completly dissociated.

Reaction 1 (dissociation) is endothermic and proceeds at high temperature. The liberated iodine is diffused back to the impure metal which is maintained at lower temperature and the reverse reaction occurs (reaction 2). The cycle is repeated until purification is complete. Similar systems for the purification of titanium, zirconium and other metals have been proposed as reviewed in reference (1).

Another well known closed-reactor process is pack cementation where the driving mechanism for the reaction is not a temperature difference but a difference in chemical activity in an isothermal system. It is a very useful process which is used industrially on a large scale for chromizing, aluminizing and siliconizing, as shown schematically in Figure 1 (1,2). Although the process unquestionably involves vapor deposition reactions locally, these reactions are not controllable once the pack constituents and operating temperature have been chosen, thus pack cementation will not be considered in any detail here.

In open-reactor CVD or flowing gas CVD (as opposed to closed reactor CVD), the reactants are introduced continuously and flow through the reactor. Closed-reactor CVD is by far the most common system and will be referred to simply as CVD. It comprises three interrelated components: a) the reactant supply, b) the deposition system or reactor (the term reactor is universally used in CVD parlance to describe the vessel in which the reaction takes place and has nothing to do with a nuclear reactor) and c) the exhaust system. A typical CVD system is shown schematically in Figure 2.

3.0 REACTANT SUPPLY

3.1 Reactant Transport

The reactants (including diluent, extender and carrier gases) must be transported and metered in a controlled manner into the reactor. In the case of gaseous reactants, this does not present any particular problem and is accomplished by means of pressure gauges and flowmeters, usually mass or ball type flowmeters.

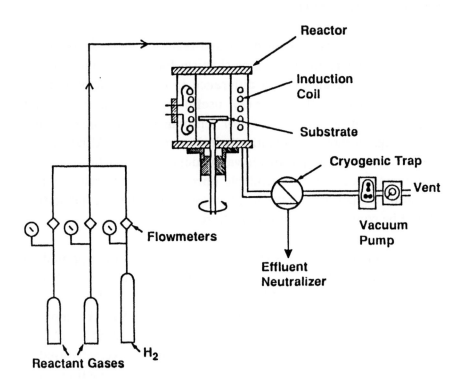

Figure 2. Schematic Diagram of Typical CVD Apparatus

Reactants that are liquid at room temperature must be heated to their evaporation temperature and transported into the reaction chamber by a carrier gas which may be an inert gas such as argon or another reactant such as hydrogen. If the vapor pressure of the liquid reactant is known, its partial pressure can be calculated and controlled by controlling the rate of flow and the volume of the carrier gas. Another vaporizer design uses flash vaporization where the liquid is metered into a vessel heated above its boiling point (at the prevailing pressure). Controlled metering of the liquid into the vaporizer can be accomplished by a peristaltic pump or similar devices. A typical vaporizer to supply $TiCl_4$ is shown in Figure 3.

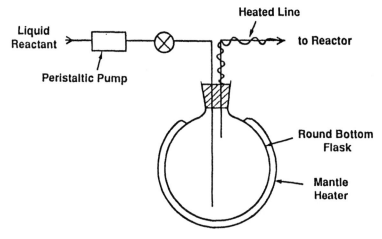

Figure 3. Vaporizer for Liquid Reactant

Reactants which are solid at room temperature present more of a problem since they must be heated to their vaporization temperature which in some cases may be relatively high. This is particularly true with metal halides as shown in Chapter Three, Table 3. It is often preferable to generate the reactant *in situ* in a manner similar to the purification of chromium mentioned in Section Two above. An example is the deposition of tungsten. Tungsten metal in the form of chips, pellets or powder is placed in a chlorinator shown in Figure 4 where the following reaction occurs:

[3] $W + 3Cl_2 \longrightarrow WCl_6$
                    $800°C$

The chloride is then reduced by hydrogen as follows:

[4] $WCl_6 + 3H_2 \longrightarrow W + 6HCl$

800-1200°C

The design of the chlorinator must be such that hydrogen cannot get into the chlorine supply and vice versa otherwise an explosion may occur. For that reason, chlorine is usually metered with an inert gas such as argon and the hydrogen is introduced below the chlorinator.

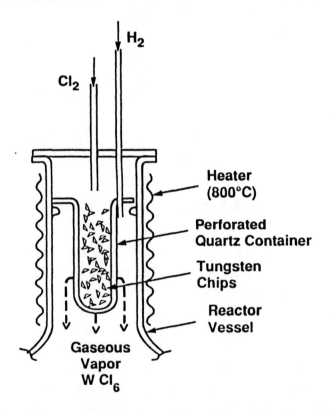

Figure 4. *In Situ* Chlorinator for the Generation of Tungsten

More hydrogen is required in the lower range of temperature and gradually less as the temperature is increased. Above 1000°C, WCl_6 decomposes without the need for any hydrogen.

The amount of tungsten chloride generated in Reaction 3 is a function of the surface area of the tungsten chips and will be gradually reduced as the reaction proceeds and will eventually stop when all the metal has been chlorinated.

3.2 Reactant Purity and Contamination

The requirements placed on the performance and reliability of CVD coatings are continuously upgraded in every area of application. This means an ever increasing degree of purity of the precursor materials since impurities are the major source of defects in the deposit. The purity of a gas is expressed in terms of nines, for instance six nines, meaning a gas that is 99.9999% pure which is now a common requirement. It is also expressed in ppm (parts per million) or ppb (parts per billion) of impurity content.

Suppliers of precursor materials are well aware of this problem and are making considerable efforts to improve their products, usually at a greatly increased cost. However using a pure reactant is not enough since a gas can become contaminated again as it leaves its storage container and travels through the distribution system to the reactor chamber. It can pick up moisture, oxygen, particles and other contaminants even if gastight metal lines are used. For that reason, it is recommended to purify and filter the gases at the point of use, i.e. just before entering the reaction zone (3, 4).

There are several methods used to purify gases: catalytic adsorption, palladium diffusion, gettering, chemisorption and filtration (5).

Catalytic Adsorption: This method can reduce impurities such as H_2, O_2, CO and hydrocarbons to less than 10 ppb. The catalyst converts these impurities into CO_2, H_2O and other species that can then be removed by molecular sieves and cryogenic adsorption.

Palladium Diffusion: Palladium is very permeable to hydrogen but not permeable to other gases. As a result, it is a useful hydrogen purifier. A palladium membrane, heated to 400°C, purifies hydrogen to < 10ppb but requires a high pressure differential for net diffusion to take place at reasonable rates of hydrogen supply.

Gettering: Gettering materials, such as zirconium or titanium alloys, are heated to 400°C. At that temperature,they react with the impurities in the gas stream such as O_2, H_2O , N_2, H_2, CO, CO_2 and hydrocarbons. Total impurities can be reduced to < 100ppb.

Chemisorption. This process uses resins which are chemically treated with metallo-organics and discarded after depletion.

Filtering. This is the final step after purification. Polymers such as Teflon are used widely in filters but, because of problems with their outgassing, are being increasingly replaced by ceramic and metal filters.

4.0 THERMAL CVD: DEPOSITION SYSTEM AND REACTOR

As stated in the introduction to this chapter, CVD can be classified by the method used to apply the energy necessary to activate the CVD reaction, i.e. thermal, plasma, laser and photon. In this section, thermal CVD is reviewed. Thermal CVD is the oldest and is often referred to as "conventional CVD". As the name implies, thermal CVD is activated by temperature.

4.1 Heating Methods

Thermal CVD is characterized by the need for high temperature, generally from 800 to 2000°C which can be generated by resistance heating, high frequency induction, radiant heating, hot plate heating or any combination of these. Reactors can be horizontal or vertical. This makes for a sometimes bewildering proliferation of designs further complicated by the requirement for control of pressure which can range from atmospheric to a few mTorr.

Thermal CVD can be divided into two basic systems known as the hot wall reactor and the cold wall reactor.

Hot wall reactors. A hot wall reactor is essentially an isothermal furnace which is generally heated by resistance elements, usually as a batch operation. The parts to be coated are loaded, the furnace temperature is raised to the desired level and the reaction gases are introduced. Figure 5 shows such a furnace which is used for the coating of cutting tools with TiC, TiN and Ti(CN). These materials can be deposited alternatively under very precisely controlled conditions. Such reactors can be very large and the coating of hundreds of parts in one operation is possible (see Chapter Eleven). Another type of reactor, used in the coating of doped silicon on semiconductor wafers, is shown in Figure 6. The wafers are stacked vertically, which minimizes particle contamination and considerably increases the capacity (as opposed to horizontal loading). Deposition usually takes place at low pressure (e.g. 1 Torr).

Hot wall reactors have the advantage of very close temperature control. A disadvantage is that deposition occurs everywhere, on the part as well as on the walls of the reactor which require periodic cleaning or the use of a disposable liner.

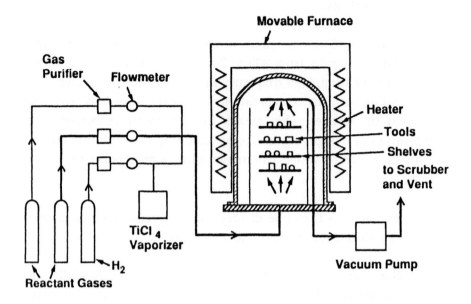

Figure 5. Production CVD Reactor for the Coating of Cutting Tools

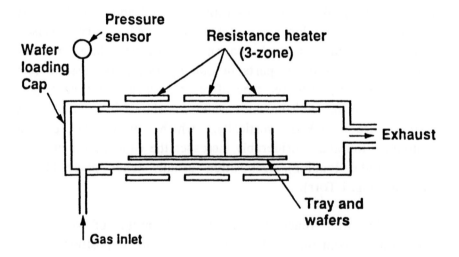

Figure 6. Low Pressure CVD Reactor for the Coating of Doped Silicon Wafers

Cold wall reactors: In a cold wall reactor (also known as an adiabatic wall reactor), the substrate to be coated is heated directly either by induction or by radiant heating. Most CVD reactions are endothermic, i.e. they absorb heat. As a result, deposition takes place preferentially on the surfaces where the temperature is the highest, in this case the substrate, while the walls of the reactor, which are cooler, remain uncoated. A simple laboratory type reactor is shown in Figure 7. It is used for the deposition of tungsten on graphite using *in situ* chlorination. It is heated by high frequency (450 KHz) induction and operates at low pressure (5 Torr). The part is rotated to improve deposition uniformity.

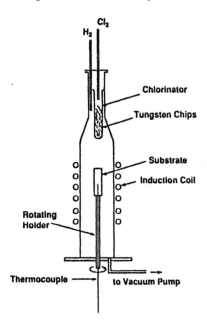

Figure 7. Cold-Wall Laboratory Reactor for Tungsten Deposition

A more elaborate example of induction heating is shown in Figure 8 which is a schematic of a twin chamber reactor used for semiconductor silicon epitaxy in VLSI, bipolar and MOS devices (see Chapter Eight). The power is supplied by a solid state high frequency (20KHz) generator. A radiation reflector, shown in Detail

A, increases the efficiency and uniformity of deposition. Pressure can be from 50 mTorr to 1 atm.

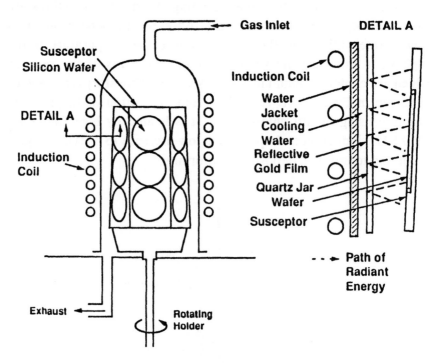

Figure 8. Schematic Diagram of Cold-Wall Production Reactor for Silicon Epitaxy

Another example of a cold wall reactor is shown in Figure 9. It uses a hot plate and a conveyor belt for continuous operation at atmospheric pressure. Preheating and cooling zones reduce the possibility of thermal shock. The system is used extensively for high volume production of silicon dioxide films for semiconductor passivation and interlayer dielectrics.

Radiant heating, which in the past has been used mostly in experimental reactors, is gradually being introduced into production systems. The basic design is shown in Figure 10. It is of course essential that the walls of the reactor be transparent to radiation and remain so during the deposition sequence so that heating can proceed unhindered.

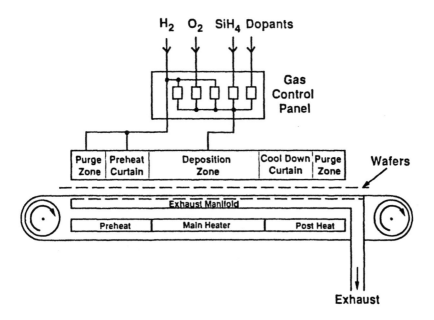

Figure 9. Continuous Operation Cold-Wall Reactor for Atmospheric Pressure Deposition of SiO_2

4.2 Atmospheric and Low Pressure Reactors

It was shown in Chapter Two that the effect of pressure on film deposition can be considerable. At atmospheric pressure, the rate of gas-phase transport (i.e. the rate of diffusion through the boundary layer) of both reactant gases (diffusing in) and by- product gases (diffusing out) is low and the reaction is diffusion limited.

At low pressure on the other hand (i.e. < 1 Torr), the rate of gas-phase transport of both reactants and by-products is increased in inverse proportion to pressure, by a factor of 100 if P is decreased from 760 to 7.6 Torr. However, one may not gain similarly if the pressure decrease is at the expense of the partial pressure of reactant gas, since e.g., the kinetic rate (for first-order reactions) may be directly proportional to the reactant partial pressure. Reduction of pressure by eliminating carrier gas is always beneficial. At low pressure, surface reaction is the rate determining step and the mass transfer variables are far less critical than at atmospheric pressure.

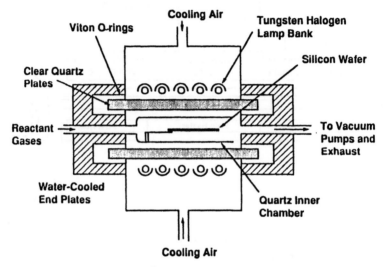

Figure 10. Cold-Wall Reactor with Radiant Heating

In practical terms, this means that low pressure generally provides films with more uniformity, better step coverage and improved quality (6). Some reactants in atmospheric pressure reactors must be highly diluted with inert gases to prevent vapor phase precipitation while generally no dilution is necessary at low pressure. However, atmospheric pressure reactors are simpler and cheaper. They can operate faster, on a continuous basis and, with recent design improvements, the quality of the deposits has been upgraded considerably, and satisfactory deposits of many materials such as oxides are obtained. The reasons for using one system or the other are not as clear-cut as they used to be.

Table 1 lists typical CVD systems used in the production of semiconductors and cutting tools. As can be seen, these systems include cold-wall and hot-wall reactors operating at low or atmospheric pressures. The decision to use a given system should be made after giving due consideration to all the factors of cost, efficiency, production rate, ease of operation and quality.

TABLE 1

Reactor Designs and Typical Production Applications

Coating Type	Application	Reactor Type	Pressure
BPSG (boro-phospho silicate glass)	passivation of semiconductors	cold wall	ca. 1 Torr
Silicon epitaxy	semiconductors to 1 atm.	cold wall	80 Torr
Silicon dioxide	passivation of semiconductors	cold wall	1 atm.
Titanium carbide and nitride	cutting tools	hot wall	1 Torr
Doped silicon	semiconductors	hot wall	1 Torr
Titania	solar cells	cold wall	1 atm.
Tungsten silicide	gates & interconnections of semiconductors	cold wall	1 Torr

A typical production reactor for the atmospheric deposition of SiO₂ and boro-phospho-silicate glass is shown in Figures 11 and 12.

Figure 11. Atmospheric Deposition Equipment for SiO$_2$ and Glass
(Source: Watkins-Johnson, Palo Alto, CA)

5.0 EXHAUST AND BY-PRODUCT DISPOSAL

CVD does not require the low pressures which are usually associated with sputtering, ion implantation and other PVD processes. Consequently the vacuum system is simpler and less costly. Mechanical pumps are adequate for most operations. Vane pumps, built with corrosion resistant material, are preferred. If properly maintained, these pumps will operate for long periods of time.

Many CVD processes use precursors which are toxic and in some cases lethal even at low concentration (for instance nickel carbonyl, diborane, arsine and phosphine). Many precursors are also pyrophoric such as silane, some alkyls, arsine and phosphine. Very often the reaction is not complete and some of the precursor materials may reach the exhaust unreacted. In addition, many of the by-products of the reaction are also toxic and corrosive. This means that all these effluents must be eliminated or neutralized before they are released to the environment. Careful design and construction of the exhaust system are essential.

Figure 12. Atmospheric CVD System Showing Silicon Wafers on Belt
(Source: Watkins-Johnson, Palo-Alto, CA)

A typical exhaust manifold is shown in Figure 13 (7). It includes a particle trap which collects solid particles originating in the reactor which might otherwise damage the pump. The trap also prevents backstreaming of pump oil into the reactor and subsequent contamination. Also in the manifold are a gate valve and a pressure indicator and sensor. If high gas velocity is required in the reactor, a roots pump is included upstream from the vane pump. Downstream from the vane pump, the exhaust gases pass through an oil demister and a scrubber where they are treated and neutralized. Also commonly used are incinerating systems for the decomposition of toxic gases such as silane, phosphine and diborane.

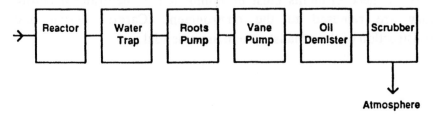

Figure 13. Schematic of Exhaust System

Exhaust conditioning requires a precise identification of all the gases and liquids, both residual precursors and by-products. It also requires leak-tight piping and ducts and suitable mechanical and chemical scrubbers to remove or neutralize dangerous materials and the proper venting of the mechanical pump with a stainless exhaust filter to remove oil mist at the source.

6.0 PLASMA CVD

In the thermal CVD process just reviewed, the reaction is activated by thermal energy and the deposition temperature may be too high for many applications. In plasma CVD, also known as plasma-enhanced or plasma-assisted CVD (PECVD or PACVD), the reaction is activated by a plasma. As a result, the substrate temperature can be considerably lower, which in many cases is a great advantage as will be discussed later.

Commercial applications of plasma CVD were developed in the 1960's, at first for semiconductor applications, notably the deposition of silicon nitride. The process has expanded ever since and is now used widely in many areas of CVD. Plasma CVD combines a chemical process with a physical process and, to some degree, bridges the gap between CVD and PVD. In this respect, it is similar to the PVD processes operating in a chemical environment such as reactive sputtering (see Appendix).

6.1 Principles of Plasma Deposition

As the temperature of a gas is increased, the atoms are gradually ionized, that is they are stripped of their electrons and a plasma is formed which consists of ions (with positive charge), electrons (with negative charge) and atoms that have not been ionized (neutral). Above some temperature, all gas molecules are dissociated into atoms which are increasingly ionized. However to accomplish this requires a very large amount of energy and, as a result, temperatures must be very high (>5000K). A plasma generated by a combustion flame has a temperature limit of approximately 3700K which is not sufficient for complete ionization. In fact such a plasma is only 10% ionized.

A more convenient way to achieve high temperature is with a low frequency discharge. By increasing the electrical energy in a fixed amount of gas, all molecules are eventually dissociated and complete ionization is achieved (8). In such a low frequency discharge, both electrons and ions respond to the constantly changing field direction; they both acquire energy and their temperature is raised more or less equally. The plasma is in equilibrium (isothermal). Isothermal plasmas are generated at relatively high pressure (100 torr to 1 atm.) At such pressure, the mean free path, that is the average distance traveled between collisions, is much reduced, collisions are more frequent and molecules and ions heat readily.

A typical example is the arc discharge, the characteristics of which are listed in Table 2. Such plasmas require a large amount of power and are extremely hot. For that reason, they have not been used to any extent in CVD deposition except for the very special and important case of diamond deposition, a topic which is reviewed in Chapter Six, Section 4.

TABLE 2

Characteristics of Plasmas Used in CVD

	Isothermal (equilibrium)	Non-isothermal (non-equilibrium)
Type of discharge	arc	glow
Frequency	1 MHz	50 KHz to 3.45 MHz 2.45 GHz (microwave)
Power	MW's	KW's
Flow rate	none	mg/s
Electron concentration	10^{14}/cm^3	10^9 - 10^{12}/cm^3
Pressure	100 Torr to 1 Atm.	<2 Torr
Electron temperature K	10^4	10^4
Atom temperature K	10^4	4×10^2

Another type of plasma of major interest in CVD is the non-equilibrium (or non-isothermal) plasma (glow discharge) which is generated in a gas by a high frequency electric field and operates at lower pressure than the isothermal plasma (see Table 2) (9,10,11,12). In such a plasma, the following events occur:

a) In the high frequency electric field, the gases are ionized into electrons and ions. The electrons, which are considerably lighter than other species, are quickly accelerated to high levels of energy corresponding to 5000K or higher. Since they have a very low mass, they do not appreciably raise the plasma temperature.

b) The heavier ions with their greater inertia cannot respond to the rapid changes in field direction, in contrast with their behavior in a low frequency field. As a result, their temperature and the temperature of the plasma remains low (hence the name non-isothermal plasma).

c) The high energy electrons collide with the gas molecules with resulting dissociation and generation of reactive chemical species and the initiation of the chemical reaction.

The high frequency currents used are radio frequency (RF) at 13.45 MHz and microwave (MW) frequency at 2.45 GHz. The use of these frequencies must comply with Federal regulations.

6.2 Characteristics of the Plasma CVD Process

Plasma CVD has several advantages. It is capable of forming a deposit at temperatures where, as shown in Table 3, no reaction whatsoever would take place in thermal CVD. This is probably its major advantage since it permits the coating of low-temperature substrates such as aluminum (which might otherwise melt) or organic polymers (which would otherwise degrade and outgas) or of metals or metal alloys which experience structural changes at high temperature such as austenitic steel. Dopants for semiconductors such as boron and phosphorus can also readily diffuse between buried layers of the device if the temperature exceeds 800°C, resulting in detrimental changes in the semiconductor properties. With plasma CVD these doped materials can easily be coated without diffusion.

TABLE 3

Typical Deposition Temperatures for Thermal and Plasma CVD

	Deposition Temperature (°C)	
Material	Thermal CVD	Plasma CVD
Epitaxial silicon	1000-1250	750
Polysilicon	650	200-400
Silicon nitride	900	300
Silicon dioxide	800-1100	300
Titanium carbide	900-1100	500
Titanium nitride	900-1100	500
Tungsten carbide	1000	325-525

Another advantage is that the effects of thermal expansion mismatch between substrate and coating and resulting stresses are reduced since the temperature of deposition remains low.

In addition, the deposition rates are usually increased and, since the pressure is low (in the case of glow discharge plasma CVD), the rate controlling factor is surface kinetics (see Chapter Two, Section 3.3), which leads to greater uniformity. The low temperature of deposition also favors the formation of amorphous or very fine grained polycrystalline deposits which usually have superior properties.

The limitations of plasma CVD are as follows. It is difficult to obtain a deposit of pure material. In most cases, desorption of by-products and other gases is incomplete because of the low temperature and these gases, particularly hydrogen, remain as inclusions in the deposit. Moreover, in the case of compounds such as nitrides, oxides, carbides or silicides, stoichiometry is rarely achieved. This is generally detrimental since it alters the physical properties and reduces the resistance to chemical etching and radiation attack. However in some cases, it is advantageous; for instance, amorphous silicon used in solar cells has improved opto-electronic properties if hydrogen is present (see Chapter Six, Section 5).

Plasma CVD tends to create undesirable compressive stresses in the deposit particularly at the lower frequencies (11). This may not be a problem in very thin films used in semiconductor applications, but in thicker films typical of metallurgical applications, the process is conducive to spalling and cracking.

Another disadvantage is that fragile substrates used in VLSI, such as some III-V and II-VI semiconductors materials, can be damaged by the ion bombardment from the plasma, particularly if the ion energy exceeds 20 eV. In addition, the plasma reacts strongly with the surface of the coating as it is deposited. This means that the deposition rate and often the film properties depend on the uniformity of the plasma. Areas of the substrate fully exposed will be more affected than the more sheltered ones. Finally, the equipment is generally more complicated and more expensive.

Overall the advantages of plasma CVD are considerable and it is used in an increasing number of applications as will be shown below.

6.3 Materials Deposited by Plasma CVD

The following table lists examples of plasma CVD materials and their applications (10,11,12).

TABLE 4

Examples of Plasma CVD Materials and Applications

Material Deposited	Common Precursors	Deposition Temp. °C	Applications	Status
a-Si	SiH_4-H_2	250	semiconductor photovoltaic	production
Epitaxial silicon	SiH_4	750	semiconductor	R & D
Si_3N_4	SiH_4-N_2-NH_3	300	passivation	production
SiO_2	SiH_4-N_2O	300	passivation optical fiber decorative	production

TABLE 4 (cont.)

Examples of Plasma CVD Materials and Applications

Boro-phos-pho silicate	SiH_4-TEOS-B_2H_6-PH_3	355	passivation	semi-production
W	WF_6	250-400	conductor in IC's	R & D
WSi_2	WF_6-SiH_4	230	conductor in IC's	semi-production
$TiSi_2$	$TiCl_4$-SiH_4	380-450	conductor in IC's	semi-production
TiC	$TiCl_4$-$C2H_2$	500	abrasion cutting tools	R & D
TiN	$TiCl_4$-NH_3	500	abrasion cutting tools	R & D
Diamond-like carbon	CH_4-H_2 hydrocarbon	300	wear, erosion optical	semi-production

As mentioned above, the major advantage of plasma CVD over thermal CVD is its lower deposition temperature. This is particularly important in the production of the new semiconductor devices such as VLSI's with film thicknesses in the order of 2000 Å which means that the possibility of interdiffusion is high and high processing temperature are not acceptable. Many IC designs also use aluminum and the temperature cannot exceed 500°C. As a result, semiconductor applications were the first to use plasma CVD and they remain the major user of the technology. It is only recently that it has been considered for other applications such as wear and corrosion protection and cutting tools.

The following materials are the major ones deposited by plasma CVD. Their deposition is described in greater detail in Chapter Seven.

Silicon Nitride: Si_3N_4 forms an excellent passivation film since it is somewhat elastic and does not crack. It provides an effective barrier against water and sodium ions. In addition, its adhesion to aluminum and gold is good. Stoichiometry is difficult to control and hydrogen usually remains incorporated in the lattice. The material then is best described as $Si_xN_yH_z$.

Silicon Dioxide: SiO_2 by plasma CVD is used extensively as an insulator in semiconductor devices and, more recently, as a coating for optical fibers and in some decorative applications. The preferred precursor materials are silane (SiH_4) and nitrous oxide (N_2O).

Borophosphosilicate Glass: This type of glass is deposited by plasma CVD for the passivation of semiconductor devices using silane, tetraethyl orthosilicate (TEOS), N_2O, O_2 or CO_2.

6.4 Plasma CVD Equipment

Plasma CVD equipment for metering and measuring gas flow, temperature and pressure is similar to that for conventional CVD. So is the exhaust system. The difference is in the design of the reactor and the equipment necessary to produce the plasma.

Most production plasma reactors use radio frequency (RF) or microwave (MW) glow discharge with parallel electrodes. A typical reactor is shown in Figure 14. It is a cold wall design which is used on a large scale for the deposition of silicon nitride and silicon dioxide for semiconductor devices. There are several variations of this basic design. A typical reactor operates at 450 KHz or 113.56

MHz (11).

A recent and promising development uses electron cyclotron resonance (ECR) to produce the plasma, by the proper combination of an electric field and a magnetic field. Cyclotron resonance is achieved when the frequency of the alternating electric field is made to match the natural frequency of the electrons orbiting the lines of force of the magnetic field. This occurs for instance with a frequency of 2.45 GHz (a standard frequency for microwave) and a magnetic field of 875 Gauss. An ECR plasma reactor is shown schematically in Figure 15 (11,13,14).

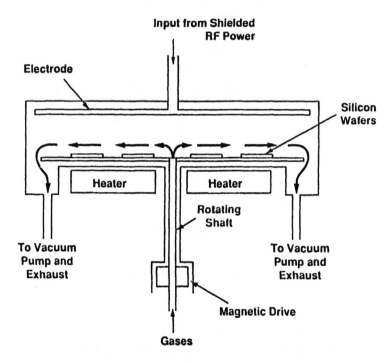

Figure 14. Radial Flow RF Plasma Reactor

An ECR plasma has two basic advantages. The possibility of damaging the substrate by high intensity ion bombardment is minimized. In an RF plasma reactor such as the one shown in

Figure 14, the ion energy may reach 100eV which could easily damage devices having submicron line width features and made from the more fragile compound semiconductor materials such as gallium arsenide, indium phosphide, mercury cadmium telluride or gallium aluminum arsenide (III-V and II-VI compounds).

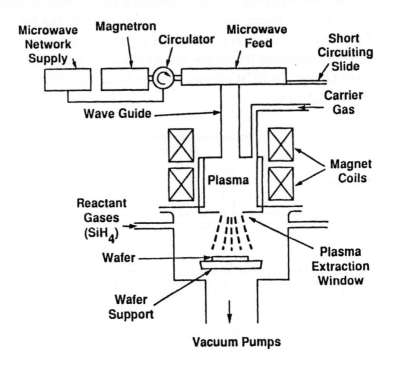

Figure 15. Schematic Diagram of ECR Microwave Deposition Apparatus

Another advantage of the ECR plasma is that it operates at lower temperature than an RF plasma and the risk of damaging heat-sensitive substrate is minimized. Also minimized is the possibility of forming hillocks, which are the result of the recrystallization of the conductor metals, particularly aluminum. Hillock formation often occurs when the aluminum is coated with a dielectric if the deposition temperature exceeds 400°C. Deposition below that temperature is easily achieved with an ECR plasma especially in the case of SiO_2 which can be deposited as low as 300°C.

A limitation of ECR systems is the need for very low pressure (10^{-3} to 10^{-5} Torr) as opposed to 0.1 to 1 Torr for RF plasma systems as well as the need for a high intensity magnetic field. This means more costly equipment. In addition, since there is the added variable of the magnetic field, the processing is more difficult to control.

7.0 LASER AND PHOTO CVD

Thermal and plasma activation are the major systems used to activate a CVD reaction. Two others methods have recently been developed: laser and photo CVD. These methods are still essentially in the experimental stage, but have a great deal of potential at least in specialized areas.

7.1 Laser CVD

A laser produces a coherent, monochromatic, high energy beam of photons which can be used effectively to activate a CVD reaction. This activation can take place by two different mechanisms: a) thermal and b) photolytical (15).

Thermal laser CVD (also known as laser pyrolysis) occurs as a result of the thermal energy from the laser coming in contact with and heating an absorbing substrate. The wavelength of the laser is such that little or no energy is absorbed by the gas molecules. The substrate is locally heated in a manner analogous to the local heating in a cold-wall reactor and deposition is restricted to the heated area as shown in **Figure 16** which illustrates the deposition of a thin stripe.

Thermal laser CVD involves essentially the same deposition mechanism and chemistry as conventional thermal CVD and theoretically the same wide range of materials can be deposited. The transient nature of the heating cycle can sometimes be used to advantage with heat-sensitive substrates.

Some examples of materials deposited by laser CVD are listed in Table 5.

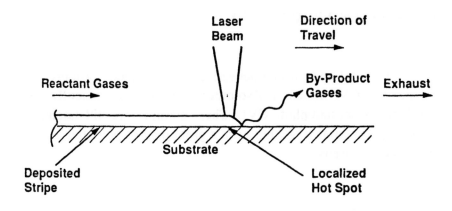

Figure 16. Schematic Diagram of Laser CVD Growth Mechanism (Stripe Deposition)

The major use so far of thermal laser CVD has been in the direct writing of thin films in semiconductor applications, using holographic methods to deposit complete patterns in a single step with width as small as 0.5 micron. Laser CVD is also used on a development basis to produce rods and coreless boron and silicon carbide fibers (see Chapter Thirteen).

TABLE 5

Examples of Materials Deposited by Thermal Laser CVD (15,16,17)

Materials	Reactants	Pressure	Laser(nm)
Aluminum	$Al_2(CH_3)_6$	10 Torr	Kr (476-647)
Carbon	C_2H_2, C_2H_6, CH_4	-	Ar-KrR488-647)
Cadmium	$Cd(CH_3)_2$	10 Torr	Kr (476-647)
Gallium Arsenide	$Ga(CH_3)_3$, AsH_3	-	Nd: YAG
Gold	Au(ac.ac.)	1 Torr	Ar
Indium oxide	$(CH_3)_3In$, O_2	-	ArF
Nickel	$Ni(CO)_4$	350 Torr	Kr (476-647)
Platinum	$Pt(CF(CF_3COCHCOCF_3))_2$	-	Ar
Silicon	SiH_4, Si_2H_6	1 atm.	Ar-Kr (488-647)
Silicon oxide	SiH_4, N_2O	1 atm.	Kr (531)
Tin	$Sn(CH_3)_4$	-	Ar
Tin oxide	$(CH_3)_2SnCl_2$, O_2	1 atm.	CO_2
Tungsten	WF_6, H_2	1 atm.	Kr (476-531)
$YBa_2Cu_3O_x$	halides	-	Excimer

7.2 Photo CVD

In photolytic CVD (photo CVD), the chemical reaction is induced   by the action of light (single photon absorption), specifically ultraviolet (UV) radiation, which has sufficient photon energy to break the chemical bonds in the reactant molecules. In many cases, these molecules have a broad electronic absorption band and they are readily excited by UV radiation. Although UV lamps have been used, more energy can be obtained from UV lasers such as the excimer lasers which have photon energy ranging from 3.4 eV (XeF laser) to 6.4 eV (ArF laser). A typical photo laser CVD system is shown schematically in Figure 17 (16,17).

Photo CVD differs from thermal laser CVD in that, since the reaction is photon activated, no heat is required and the deposition may occur essentially at room temperature. Moreover, there is no constraint on the type of substrate, which can be opaque, absorbent or transparent. Materials deposited include SiO_2, Si_3N_4, Al_2O_3, non-silicon oxides, aluminum, tungsten and other metals.

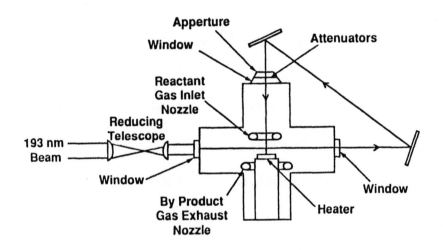

Figure 17. Schematic of Photo Laser CVD

A limitation of photo CVD is the slow rate of deposition which has so far restricted its applications. If higher power excimer lasers can be made available economically, the process could well compete with thermal CVD and thermal laser CVD, particularly in very critical semiconductor applications where low temperature is essential.

8.0 METALLO-ORGANIC CVD (MOCVD)

MOCVD is a specialized area of CVD which utilizes metallo- organic compounds as precursors usually in combination with hydrides or other reactants. The chemistry of metallo-organic compounds is reviewed in Chapter Three, Section 5. The thermodynamic and kinetic principles of CVD and its general chemistry apply to MOCVD as well.

Although the deposition of semiconductor materials by MOCVD in a closed reactor system at low pressure was first reported in 1960, it was not until 1968 that the first deposition of crystalline gallium arsenide was obtained in an open tube reaction.

These early results demonstrated that deposition of critical semiconductor materials could be obtained at lower temperature than conventional thermal CVD in a continuous flow process and that epitaxial growth could be successfully achieved. The quality of the equipment and the diversity and purity of the precursor chemicals have steadily improved since then and the process is being refined constantly (12).

A wide variety of materials can be deposited by MOCVD, either as single crystal, polycrystalline or amorphous films (18). The most important application is for the deposition of the group III-V semiconductor compounds such as gallium arsenide (GaAs), indium arsenide (InAs), indium phosphide (InP) and gallium aluminum phosphide (GaAlP), particularly for epitaxial deposition. The most commonly used precursors are trimethyl gallium (TMGa), trimethyl aluminum (TMAl), trimethyl indium (TMIn), trimethyl arsenic (TMAs) and the hydrides, phosphine (PH_3) and arsine (AsH_3). Characteristics and properties of these compounds are listed in Chapter Three, Tables 7 and 8. Safety considerations are very important in handling these compounds. A typical CVD reaction is:

[1] $(CH_3)_3Ga + AsH_3 \longrightarrow GaAs + 3CH_4$

$$600\text{-}800°C$$

Another important group of MOCVD materials consists of the II-VI compounds such as zinc sulfide (ZnS), zinc selenide (ZnSe), cadmium sulfide (CdS) and cadmium selenide (CdSe). Commonly used precursors are dimethyl cadmium (DMCd), diethyl telluride (DETe), diethyl zinc (DEZn) and hydrogen selenide (H_2Se). A typical reaction is:

[2] $(CH_3)_2Cd + (CH3)_2Te \longrightarrow CdTe + 2C_2H_6$

Most MOCVD reactions occur in the temperature range of 600 - 1000°C and at pressures varying from 1 Torr to atmospheric. A typical reactor for the deposition of GaAlAs is shown in Figure 18. Because of the highly critical requirements of most semiconductor applications, it is necessary to use the most precise equipment and extremely high purity gases (see Section 3). Electronic mass flow controllers, ultrafast bellow-sealed gas-switching valves, accurate venting control and elimination of dead space allow extremely rapid switching of gases (19). As a result, extremely thin deposits can be produced (< 100 Angstroms) with abrupt interfaces (<10 Angstroms).

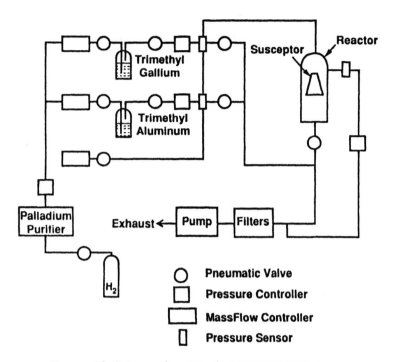

Figure 18. Schematic of Typical MOCVD System

MOCVD is presently used very extensively in microwave and opto-electronics applications, in advanced laser designs such as double heterostructure, quantum well and large optical cavity lasers as well as bipolar, field-effect transistors, infra-red detectors and solar cells (20). These applications are reviewed in Chapters Eight and Nine.

The equipment and chemicals used in MOCVD are expensive and production cost is high. For that reason, MOCVD is considered most often where very high quality is required. It has recently been investigated for other applications in the area of very high temperature oxidation protection up to 2200°C.

An example is the deposition of iridium on a mandrel with iridium ac-ac as precursor material. The iridium is then coated with rhenium and the mandrel etched away. Figure 19 shows the finished part at left and the coated mandrel at right. The final product is a spacecraft-satellite maneuvering thruster nozzle (21).

Figure 19. Deposition Mandrel for Iridium Coating
Source: Ultramet, Pacoima CA

9.0 CHEMICAL VAPOR INFILTRATION (CVI)

CVI is a special CVD process in which the gaseous reactants penetrate (or infiltrate) a porous structure which acts as a substrate and which can be an inorganic open foam or a fibrous mat or weave. The deposition occurs on the fiber (or the foam) and the structure is gradually densified to form a composite (22). The chemistry and thermodynamics of CVI are essentially the same as CVD; but the kinetics is different since the reactants have to diffuse inward through the porous structure and the by-products have to diffuse out (23). Thus maximum penetration and degree of densification are attained in the kinetically limited low-temperature regime.

CVI is used to produce reinforced metal or ceramic composites and, in this respect, competes with standard ceramic and metallurgical processes such as hot pressing and hot isostatic pressing. Such processes use high pressure and temperature and may cause mechanical and chemical damage to the substrate and its interface with the matrix. CVI, on the other hand, operates at low pressure and at temperatures lower than those required for the sintering of ceramics or the melting of metals, and the potential damage is considerably reduced.

The major limitation of CVI is the necessity of interdiffusion of reactants and reaction products through relatively long, narrow, and sometimes tortuous channels. To avoid rapid deposition and choking of the entrance end of the channels, conditions are chosen to ensure deposition in the kinetically limited regime. This is a slow process which may take several weeks before densification is achieved. Another limitation is that full densification is almost impossible to obtain due to the formation of closed porosity.

Several infiltration procedures have been developed which are shown schematically in Figure 20 (24). In isothermal infiltration (20 A), the gases surround the porous substrate and enter by diffusion. The concentration of reactants is higher toward the outside of the porous substrate, and deposition occurs preferentially in the outer portions forming a skin which impedes further infiltration. It is often necessary to interrupt the process and remove the skin by machining so that the interior of the substrate may be densified.

In spite of this limitation isothermal infiltration is used widely because it lends itself well to simultaneous processing of a great number of parts in very large furnaces. It is used for the fabrication of carbon-carbon composites for aircraft brakes and silicon carbide composites for aerospace applications (see Chapters Six and Twelve).

The principle of thermal gradient infiltration is illustrated in Figure 20 B. The porous structure is heated on one side only. The gaseous reactants diffuse from the cold side and deposition occurs only in the hot zone. Infiltration then proceeds from the hot surface toward the cold surface. There is no need to machine any skin and densification can be almost complete. Although the process is very slow since diffusion is the controlling factor, it has been used extensively for the fabrication of carbon-carbon composites including large reentry nose cones (25).

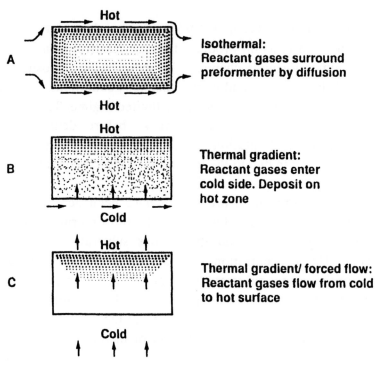

Figure 20. Chemical Vapor Infiltration (CVI) Procedures

The principle of forced flow, thermal gradient infiltration is illustrated in Figure 20 C and Figure 21 (26, 27). The gas inlet and the substrate are water cooled and only the top of the substrate is heated. The gaseous precursors enter under pressure (approximately 2 MPa back pressure) into the cool side of the substrate and flow through it to reach the hot zone where the deposition reaction takes place. The time to reach full densification is considerably reduced and final density is more uniform.

This process is well suited for the fabrication of high strength silicon carbide ceramic composites but is limited to substrates of simple configuration and is difficult to adapt to the processing of multiple units.

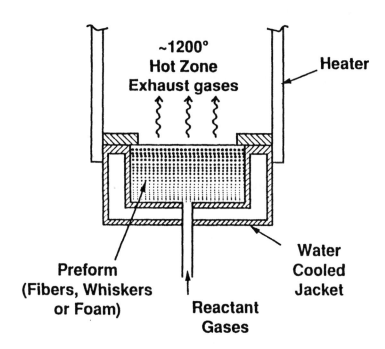

Figure 21. Forced Flow Thermal Gradient Chemical Vapor Infiltration

10.0 FLUIDIZED-BED CVD

Fluidized bed CVD is as special technique which is used primarily in the coating of powders. The powder is given quasi- fluid properties in a flowing gas. A typical fluidized bed CVD reactor is shown schematically in Figure 22.

To obtain fluidization, a careful balance must be made between the density and size of the particles to be coated and the velocity, density and viscosity of the fluidizing gas (28). The minimum gas velocity V_m (below which the particles will fall in the gas inlet and no fluidization will be obtained) is given by the following relationship:

$$V_m = d^2 \, \rho_p \cdot \rho_g \, G/1650u \; [\text{for } dV_o\rho_g/u < 20]$$

d = particle diameter

ρ_p = particle density

ρ_g = gas density

G = acceleration of gravity

V_o = superficial gas velocity

u = gas velocity

As can be seen, two factors are critical: a) the density of the particle since heavier particles are more difficult to fluidize, and b) particle size, since the necessary gas velocity varies as the square of the particle diameter.

The design of the reactor is also critical as gas velocity at the top must be less than the terminal velocity of the particles, otherwise they would be blown out of the bed.

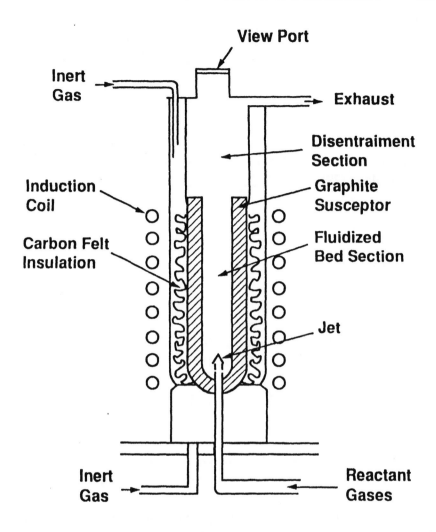

Figure 22. Fluidized Bed CVD Reactor

Fluidized-bed CVD was developed in the late fifties for a specific application: the coating of nuclear fuel particles for high temperature gas-cooled reactors. The particles are uranium-thorium carbide coated with pyrolytic carbon and silicon carbide for the purpose of containing the products of nuclear fission. The carbon is obtained from the decomposition of propane (C_3H_8) or propylene (C_3H_6) at 1350°C or methane (CH_4) at 1800°C. Methyltrichlorosilane (CH_3SiCl_3) is the preferred precursor for silicon carbide. Other compounds have been deposited by fluidized bed CVD including zirconium carbide (from $ZrCl_4$ and a hydrocarbon), hafnium carbide (from $HfCl_4$ and methane or propylene) and titanium carbide (from $TiCl_3$ and propylene). Fluidized-bed CVD is also useful in the production of high purity materials such as silicon produced by the reduction of trichlorosilane (29).

REFERENCES

1. Powell, C.F., Oxley, J.H. and Blocher, J.M. Jr., *Vapor Deposition*, John Wiley & Sons Inc., New York (1966)

2. Wachtell, R.L. and Seelig, R.P., *U.S. Patent 3,037,883* (June 5, 1962)

3. Smoak, B.C. Jr. and O'Ferrell, D., Gas Control Improves Epi Yield, *Semiconductor International*, 88-92 (June 1990)

4. Pieter Burggraaf, CVD Purity: Gas and Tool Concerns, *Semiconductor International*, pp.70-75 (March 1990)

5. Toy, D.A., Purifying at the Point of Use Keeps Your Gases at Their Cleanest, *Semiconductor International*, 64-72 (June 1990)

6. Kern, W. and Ban, V., Chemical Vapor Deposition of Inorganic Thin Films, in *Thin Film Processes*, (J. Vossen and W. Kern, Eds.), Academic Press, New York (1978)

7. Danielson, D., Yau, L. and David, K., Dry Scrub to Improve Hot Wall Nitride Equipment Reliability, *Semiconductor International*, 170-173 (Sept. 1989)

8. Thorpe, M., Plasma Energy, The Ultimate in Heat Transfer, *Chem. Eng. Progress*, 43-53, (July 1989)

9. Hollahan, J.R. and Rosler, R.S., Plasma Deposition of Inorganic Thin Films, in *Thin Film Processes*, (J. Vossen and W. Kern, Eds.), Academic Press, New York (1978)

10. Bachman P.K., Gartner, G. and Lydtin, H., Plasma- Assisted CVD Processes, *MRS Bulletin*, 51-59, (Dec. 1988)

11. Sherman, A., *Chemical Vapor Deposition for Microelectronics*, Noyes Publications, Park Ridge, NJ (1987).

12. *Handbook of Thin-film Deposition Processes and Techniques*, (K. K. Schuegraf, Ed.), Noyes Publications, Park Ridge, NJ (1988)

13. Kearney, K., ECR Finds Applications in CVD, *Semiconductor International*, 66-68, (March 1989)

14. Hu, Y.Z., Li, M., Sunko, J., Andrews, J. and Irene, E., Characterization of Microwave ECR Plasma for Low Temperature Chemical Vapor Deposition, *Proc. 11th. Int. Conf. on CVD*, (K. Spears and G. Cullen, Eds.), 166-172, Electrochem. Soc., Pennington NJ 08534 (1990)

15. Bauerle, D., Laser Induced Chemical Vapor Deposition, in *Laser Processing and Diagnostics*, (D. Bauerle, Ed.), Springer-Verlag, New York (1984)

16. Deutsch, T.F., Applications of Excimer Lasers to Semiconductor Processing, *in Laser Processing and Diagnostics*, (D. Bauerle, Ed.),

124   CHEMICAL VAPOR DEPOSITION

Springer-Verlag, New York (1984)

17. Solanki, R., Moore, C.A. and Collins, G.J., Laser-Induced CVD, *Solid State Technology*, 220-227, (June 1985)

18. Ghandhi, S.K. and Bhat, I.B., Organometallic Vapor Phase Epitaxy, *MRS Bulletin*, 37-43, (Nov. 1988)

19. Emanuel, M., *Metalorganic Chemical Vapor Deposition for the Heterostructure Hot Electron Diode*, Noyes Data Corp., Park Ridge, NJ (1989)

20. *Proc. 3d Int. Conf. on Metallo-organic Vapor Phase Epitaxy*, American Ass. of Crystal Growth, Universal City, CA (1986)

21. Harding, J.T., Tuffias, R.H. and Kaplan, R.B., *Oxidation Resistance of CVD Coatings*, AFRPL TR-86-099, Air Force Rocket Propulsion Laboratory, Edwards AFB, CA (1987)

22. Lackey, W.J., Review, Status and Future of the CVI Process for Fabrication of Fiber-Reinforced Ceramic Composites, *Ceram. Eng. Sci. Proc.*, 10(7-8):577-584 (1989)

23. Starr, T.L., Deposition Kinetics in Forced Flow/Thermal Gradient CVI, *Ceram. Eng. Sci. Proc.*, 9(7-8):803-812 (1988)

24. Besmann, T.M., Lowden, R., Sheldon, B. and Stinton, D., Chemical Vapor Infiltration, *Proc. 11th Int. Conf. on CVD*, (K. Spear and G. Cullen, Eds.) 482-491, Electrochem. Soc., Pennington, NJ 08534 (1990)

25. Pierson, H.O. amd Lieberman, M.L., The Chemical Vapor Deposition of Carbon on Carbon Fibers, *Carbon*, 13 pp. 159-166 (1975)

26. Besmann, T.M., Lowden, R.A., Stinton, D.P. and Starr, T.L., A Method for Rapid Chemical Vapor Infiltration of Ceramic Composites, *J. Phys. Collq.*, C5 Suppl.5: 50, (May 1989)

27. Tai, N. and Chou, T., Modeling of an Improved Chemical Vapor Infiltration Process for Ceramic Composites Fabrication, *J. Am. Ceram. Soc.*, 73(6):1489-98 (1990)

28. Kaae, J.L., Codeposition of Compounds by Chemical Vapor Deposition in a Fluidized Bed of Particles, *Ceram. Eng. Sci. Proc.*, 9(9-10):1159-1168 (1988)

29. Silicon is produced by the fluidized-bed reaction by Texas Instruments, Sherman, TX and Osaka Titanium, Osaka, Japan.

5

THE CVD OF METALS

1.0 INTRODUCTION

The next three chapters are a review of the materials that are deposited by CVD, either on a production basis or experimentally. The listing is not all inclusive, as the CVD of many elements and compounds has yet to be investigated or adequately reported in the literature, at least at this stage of the technology.

Each material is listed with some of its basic properties, the major CVD reactions and processes used to produce it, and its present and potential applications.

Note: The properties may vary considerably depending on processing conditions, composition and other variables, and the values listed should be viewed accordingly.

The present chapter is concerned with the CVD of metals and some metal alloys and intermetallics. The range of applications is very extensive as these metals play a very important part in the fabrication of integrated circuits and other semiconductor devices, in opto-electronic and optical applications, in corrosion protection, and in structural parts.

2.0 ALUMINUM

Aluminum is a light metal with a density of 2.69 g/cm^3, a face centered cubic structure (f.c.c), a low melting point (646°C), a high thermal expansion (23.5 ppm/°C at 25°C), high thermal conductivity (2.37 W/cm. at 25°C) and low electrical resistivity (2.65 μohm-cm at 20°C). It has relatively low strength but is highly ductile. Until recently, it was mostly deposited by evaporation or sputtering, but deposition by MOCVD is becoming popular because of the better conformal coverage it offers.

2.1 CVD Reactions

The most widely used deposition reaction is the pyrolysis of the alkyls such as:

- trimethyl aluminum, $(CH_3)_3Al$ or TMA

- triethyl aluminum, $(C_2H_5)_3Al$ or TEA

- tri-isobutyl aluminum, $(C_4H_9)_3Al$ or TIBA

- dimethyl aluminum hydride, $(CH_3)_2AlH$ or DMAH

The decomposition yields the metal and hydrocarbons. The TMA reaction has a tendency to leave carbon incorporated in the metal. Both TEA and TIBA have very low vapor pressure at room temperature and are thus difficult to use. DMAH is generally the preferred precursor. Deposition temperature range is 200-300°C and pressure up to 1 atm (1, 2). (*Note*: these alkyls are pyrophoric.)

Deposition by the hydrogen reduction of the trihalides (AlX_3) is not practical because of the stability of the latter, but is possible through the disproportionation of the monohalides. This is a difficult procedure which has been largely superseded by the decomposition of the aluminum alkyls mentioned above (3).

In addition to these low temperature thermal processes, aluminum can be deposited by the decomposition of an alkyl precursor with a UV laser (4,5,6), or with an argon-ion laser, in applications such as patterns and lithography (7,8). Aluminum has also been deposited at low temperature by magnetron plasma (9).

2.2 Applications

• Metallization of semiconductor devices and replacement of evaporated or sputtered films to improve conformal uniformity

• Coating of carbon fibers for composite fabrication

• Corrosion and oxidation protection of steel (2).

3.0 BERYLLIUM

Beryllium is a very light metal with a density of 1.85 g/cm³, an hexagonal structure, a melting point of 1277°C, a coefficient of thermal expansion of 12 ppm/°C at 25°C, a high thermal conductivity (2.01 W/cm.°C at 25°C) and an electrical resistivity of 4 μohm-cm at 20°C. Beryllium oxidizes readily and the oxide is very toxic.

3.1 CVD Reactions

The beryllium halides are very stable and consequently their reduction by hydrogen is not a practical method of obtaining the metal (3). Good beryllium deposits are obtained by the pyrolysis of the alkyls as follows (10):

$$(C_4H_9)_2Be \longrightarrow Be + H_2 + 2C_4H_8$$

The deposition temperature range is 280-305°C and the pressure is < 1 Torr. This reaction has a tendency to incorporate carbon in the deposit.

3.2 Applications

• First-wall coatings for fusion reactors (Beryllium has a low atomic number of four).

4.0 CADMIUM

Cadmium is a toxic, soft metal with good corrosion resistance, low melting point (321°C) and high thermal expansion (30 ppm/°C at 25°C). It has an hexagonal structure, a thermal conductivity of 0.91 W/cm.°C at 25°C and an electrical resistivity of 6.83 μohm-cm at 20°C. It is normally deposited by electro-plating and its processing by CVD has recently been investigated (11).

4.1 CVD Reactions

Cadmium has been deposited by thermal laser CVD using the decomposition of dimethyl cadmium, $Cd(CH_3)_2$, at 10 Torr with an argon laser (11).

5.0 CHROMIUM

Chromium is a hard metal with excellent corrosion and oxidation resistance. It has a cubic, body-centered structure, a density of 7.19 g/cm³, a thermal expansion of 6.2 ppm/°C at 25°C, a thermal conductivity of 0.91 W/cm.°C at 20°C and an electrical resistivity of 12.90 μohm-cm at 20°C. Chromium coatings are usually produced by electroplating or sputtering but CVD is being investigated for several applications.

5.1 CVD Reactions

Chromium can be deposited by the pyrolysis of the iodide, which can be prepared in situ by passing a flow of iodine vapor over the metal at 700°C. The iodide is then decomposed at 1000°C at pressures up to 1 atm. as follows:

$$Cr + 2I \longrightarrow Cr + I_2$$

Note: Iodine is mostly monatomic at 1000°C, depending on pressure.

Chromium can also be prepared by the hydrogen reduction of the chloride at 1200 to 1325°C as follows (3, 12, 13):

$$CrCl_2 + H_2 \longrightarrow Cr + 2HCl$$

Chromium can also be deposited by MOCVD by the decomposition of dicumene chromium, $(C_9H_{12})_2Cr$, at 320-545°C (14, 15). However, the reaction tends to incorporate carbon or hydrogen in the deposit. It can also be deposited by the decomposition of its carbonyl which is made by dissolving the halide in an organic solvent such as tetrahydrofuran with CO at 200-300 atm. and at temperatures up to 300°C in the presence of a reducing agent such as an electropositive metal (Na, Al or Mg), trialkylaluminum and others (3).

5.2 Applications

• Corrosion protection and oxidation protection of steels and other metals

• Experimental contact metallization in integrated electronic circuits

6.0 COPPER

Copper is a malleable and ductile metal with high thermal conductivity (3.98 W/cm°C at 25°C) and a very low electrical resistivity (1.673 μohm-cm at 20°C). It has a face centered cubic structure, a density of 8.96 g/cm³, a melting point of 1083°C and a high thermal expansion (16.6 ppm/°C at 25°C). It is generally produced by electroplating but CVD processing is being investigated to deposit electrical conductors in semiconductor applications.

6.1 CVD Reactions

Deposition can be carried out by the hydrogen reduction of the chloride at high temperature (up to 1000°C):

$$CuCl_2 + H_2 \longrightarrow Cu + 2HCl$$

Deposition occurs at much lower temperature (260-340°C) by the decomposition of copper acetylacetonate, $Cu(C_5H_7O_2)_2$, or by the hydrogen reduction of the copper chelate, $Cu(C_5HF_6O_2)_2$ at 250°C (16, 17).

6.2 Applications

- Conductive coatings for semiconductor applications (18)

- Alloying element with CVD aluminum to reduce electromigration

7.0 GOLD

Gold is an extremely malleable and ductile precious metal with excellent corrosion and oxidation resistance. It has high thermal conductivity (3.15 W/cm.°C at 25°C) and low electrical resistivity (2.35 μohm-cm at 20°C). It has high density (18.8 g/cm³), a face-centered cubic structure, a melting point of 1064°C and a thermal expansion of 14.2 ppm/°C at 25°C. It is generally deposited by electroplating or sputtering but its CVD is being considered in semiconductor applications.

7.1 CVD Reactions

Since gold has a relatively low melting point, high temperature reactions are not possible. It is deposited by the decomposition of metallo-organics such as the following (19):

- dimethyl 1-2,4 pentadionate gold (III)

- dimethyl- (1,1,1-trifluoro-2-4-pentadionate) gold (III)

- dimethyl-(1,1,1-5,5,5 hexafluoro 2-4 pentadionate) gold

(III)

In addition to thermal MOCVD, laser CVD has been successfully demonstrated (20, 21).

7.2 Applications

- Contact metallization and metallization of alumina in semiconductor applications.

8.0 IRIDIUM

Iridium is a hard and brittle metal, which makes it difficult to machine or form. This limitation has spurred the development of vapor processing. Iridium is the most corrosion and oxidation resistant metal and, with osmium, the densest element with a density of 22.4 g/cm³. It has a very high melting point (2410°C), a face centered cubic structure, a thermal conductivity of 1.47 W/cm.°C at 25°C, a thermal expansion of 6 ppm/°C at 25°C and an electrical resistivity of 5.3 μohm-cm at 20°C.

8.1 CVD Reactions

In early experiments, iridium was deposited by the hydrogen reduction of the fluoride at 775°C (22):

$$IrF_6 + 3H_2 \longrightarrow Ir + 6HF$$

This reaction does not produce a satisfactory coating.

A commonly used deposition reaction is the pyrolysis of the acetyl acetonate, $Ir(CH_3COCHCOCH_3)_3$. The vaporization of this precursor is usually obtained in a fluidized bed at 235°C. Deposition temperature range is 400-450°C and pressure is 1 atm.. Carbon tends to be incorporated in the deposit above these temperatures (23, 24). With the proper conditions, high quality, dense deposits are obtained by this method.

8.2 Applications

• Corrosion resistant and oxidation resistant coatings for rocket engines and other aerospace applications

• Coatings for thermionic cathodes

9.0 IRON

Pure iron is very reactive chemically and oxidizes readily. It has four allotropic forms, one of which (alpha) is magnetic with a Curie transition point of 770°C. It has a density of 7.6 g/cm^3, a melting point of 1536°C, a thermal expansion of 12.6 ppm/°C at 25°C, a thermal conductivity of 0.80 W/cm.°C at 25°C and an electrical resistivity of 9.71 μohm-cm at 20°C.

9.1 CVD Reaction

Iron has been deposited by the hydrogen reduction of its chloride at 650°C or the pyrolysis of its iodide at 1100°C. (3). Recent studies have shown the advantages of the classical carbonyl decomposition at 370-450°C (25):

$$Fe(CO)_5 \longrightarrow Fe + 5CO$$

CO is used as carrier. Carbon tends to be incorporated in the deposit requiring a 900°C annealing in H_2 to remove it.

9.2 Applications

- Epitaxial grown iron films on GaAs for semiconductor applications

10.0 MOLYBDENUM

Molybdenum is a refractory metal with a body-centered cubic structure, It has high strength and is easily fabricated. It has relatively low thermal expansion (5 ppm/°C at 25°C), a thermal conductivity of 1.4 W/cm.°C and a low electrical resistivity of 5.2 μohm-cm at 0°C. Molybdenum recrystallizes above 950°C with accompanying reduction in mechanical properties. CVD is commonly used for the production of molybdenum coatings and free-standing shapes.

10.1 CVD Reactions

The hydrogen reduction of the chloride is the most commonly used deposition reaction (3):

[1] $MoCl_6 + 3H_2 \longrightarrow Mo + 6HCl$

This reaction has a ΔG^O = -114 Kcal/mole Mo at 800K and it occurs over a wide range of temperature (400-1350°C). Best deposits are obtained at low pressure (<20 Torr) and at the high end of the temperature range. The reaction is used in the production of metallurgical coatings.

Another common reaction is the silicon reduction of the fluoride:

[2] $MoF_6 + 3Si \longrightarrow 2Mo + 3SiF_4$

This reaction has a ΔG^0 = -162 Kcal/mole Mo at 800K. Deposition temperature is 200-500°C at low pressures (<20 Torr). MoF_6 tends to decompose into F_2 and solid MoF_3 which results in inconsistency in the coating especially at temperatures <400°C. This reaction, which has been considered for the metallization of IC's, is now largely superseded by tungsten deposition (26, 27).

Another well established metallization reaction is the decomposition of the carbonyl:

[3] $Mo(CO)_6 \longrightarrow Mo + 6CO$

Reaction temperature ranges from 300 to 700°C and pressure from about 1 Torr to 1 atm. The reaction is carried out in a hydrogen atmosphere to reduce the possibility of carbon contamination. A deposition temperature > 450°C is required to eliminate the incorporation of C and O_2 in the deposit (28, 29, 30).

In addition to the thermal CVD systems mentioned above, molybdenum is deposited by plasma CVD using Reaction 3 in hydrogen (31). Annealing is required to remove incorporated carbon and oxygen.

10.2 Applications

- IC's contact and gate metallization using reaction 3

• Schottky contact metallization

• Erosion resistant coating for gun-steel tubes with carbonyl precursor (31)

• Coatings for photothermal solar converters with high infrared reflectance, which use Reaction 3 and a 1000°C anneal (32)

• Coatings for high power laser mirrors using reaction 3 and thermal annealing (33)

• Free-standing shapes such as tubes, rods, etc.

11.0 NICKEL

Nickel is a hard, ductile, malleable and ferromagnetic metal. It has a face centered cubic structure with a density of 8.90 g/cm³, a melting point of 1453°C, a thermal expansion of 13 ppm/°C at 25°C, a thermal conductivity of 0.899 W/cm.°C at 25°C and an electrical resistivity of 6.84 μohm-cm at 20°C. Nickel coatings are mostly produced by electroplating, but CVD nickel is gaining in importance.

11.1 CVD Reactions

The most commonly used reaction is the decomposition of nickel carbonyl:

[1] $Ni(CO)_4 \longrightarrow Ni + 4CO$

This reaction can occur over a temperature range of 150-200°C with optimum temperature of 180-200°C. Pressure is up to 1 atm. (see Chapter Three, Section 4.2). A partial pressure of CO inhibits the reaction. Above 200°C, carbon tends to deposit together with the nickel. The reaction can be catalyzed by hydrogen sulfide and deposition can be obtained as low as 100°C (3,35,36). More ductile nickel is obtained at low deposition rates (*Note:* Nickel carbonyl is highly toxic.) $Ni(CO)_4$ is obtained by reacting finely divided nickel powder with CO. This reaction proceeds at atmospheric pressure and at temperatures between 45 and 90°C with optimum temperature of 80°C. It occurs in two steps as follows:

[2] $Ni + O_2 \longrightarrow NiO_2$

[3] $NiO_2 + 6CO \longrightarrow Ni(CO)_4 + 2CO_2$

This is the Mond reaction, first developed by Ludwig Mond in 1890. It is still used for the production and purification of metallic nickel. A similar reaction is used for the production of $Fe(CO)_5$ but higher temperature is required.

Nickel is also deposited from the alkyl at 200°C as follows (37):

$Ni(C_5H_5)_2 + H_2 \longrightarrow Ni + 2C_5H_6$

It can also be deposited by the hydrogen reduction of the nickel chelate, $Ni(C_5HF_6O_2)_2$ at 250°C (38). In addition to thermal processing reviewed above, nickel is deposited by laser CVD from the carbonyl with a krypton or a pulsed CO_2 laser(39,40).

11.2 Applications

• Molds, dies and other forming tools for metal and plastic processing, especially those involving irregular surfaces and internal areas

• High strength structural parts when alloyed with small amount of boron (0.05-0.2 wt.%) produced by codeposition from $Ni(CO)_4$ and B_2H_6 (41)

• Contacts for electronic applications (alloyed with palladium)

12.0 NIOBIUM (COLOMBIUM)

Niobium is a soft, ductile, refractory metal with high strength retention at high temperature, a relatively low density (8.57 g/cm^3), a high melting point (2468°C) and a low capture cross-section for thermal neutrons. It has a body-centered cubic structure, a thermal conductivity of 0.52 W/cm.°C at 25°C, a thermal expansion of 7 ppm/°C at 25°C and an electrical resistivity of 12.5 μohm-cm at 0°C. It is readily attacked by oxygen and other elements above 200°C. CVD is used to produce coatings or free standing shapes.

12.1 CVD Reactions

Niobium is generally deposited by the hydrogen reduction of its chloride which is accomplished at 900-1300°C and at pressure up to 1 atm. as follows:

[1] $NbCl_5 + 2\frac{1}{2} H_2 \longrightarrow Nb + 5HCl$

The chloride is usually prepared by direct chlorination *in situ* by heating the metal chips to 300°C (see Chapter 4, Section 3.1).

Deposition is also feasible by the hydrogen reduction of the bromide at 1200°C as follows:

[2] $NbBr_5 + 2\frac{1}{2} H_2 \longrightarrow Nb + 5HBr$

If a graphite substrate is used, Reaction 2 produces niobium carbide which may remain incorporated in the metal deposit (42). Niobium absorbs hydrogen readily and, for that reason, an inert atmosphere such as argon is used in the heating and cooling cycles.

12.2 Applications

- Coatings for nuclear fuel particles

- Cladding for steel and copper tubing for chemical processes

13.0 PLATINUM

Pure platinum is malleable and ductile with excellent corrosion resistance. It has very high density (21.45 g/cm³), a face-centered cubic structure, a thermal expansion of 9 ppm/°C at 25°C (identical to soda-lime glass), a thermal conductivity of 0.73 W/cm.°C at 25°C and an electrical resistivity of 10.6 μohm-cm at 20°C. When alloyed with cobalt, it has excellent magnetic properties (76.7 w% Pt, 23.3 w% Co).

13.1 CVD Reactions

The platinum halides are volatile with a decomposition point too close to the vaporization point to make them practical for CVD transport (3). Platinum can be produced from the decomposition of the acetyl acetonate, although carbonaceous impurities remain in the deposit. The carbonyl halides, specifically the dicarbonyl dichloride, are more satisfactory precursors. The reaction is as follows (presumed decomposition):

$$Pt(CO)_2Cl_2 \longrightarrow Pt + 2CO + Cl_2$$

Reaction temperature range is 500-600°C. The carbonyl partial pressure is 10-20 Torr (43).

Platinum produced by the decomposition of the complex: platinum hexafluoro-2,4-pentadionate, $Pt(CF_3COCHCOCF_3)_2$, by an argon laser has also been reported (44). The metal can also be obtained from the decomposition of tetrakis-trifluorophosphine, $Pt(PF_3)_4$ at 200-300°C in an atmosphere of hydrogen (43).

13.2 Applications

- Coatings for high temperature crucibles

- Catalyst in fuel cells and automobile emission control

- Ohmic and Schottky diode contacts

- Diffusion barrier metallization

14.0 RHENIUM

Rhenium is a refractory metal with a very high melting point (3180°C), second only to tungsten among the metals. It is far less brittle than tungsten. It has a very high modulus of elasticity and it maintains its h.c.p. crystal structure to its melting point. It has a very high rate of work hardening. Annealed rhenium is very ductile and can be bent, coiled or rolled. At elevated temperature, its ultimate tensile strength is considerably higher than that of the other refractory metals, including tungsten, and is still appreciable above 2000°C.

Rhenium is generally very chemically resistant due to its position in the periodic table next to the noble metals of the platinum group. However its major drawback is its rapid degradation in oxidizing environment and, in this respect, it is less resistant than the other refractory metals. It has very high density (21 g/cm³), a thermal conductivity of 0.48 W/cm.°C at 20°C, an electrical resistivity of 19.3 μohm-cm at 20°C and a thermal expansion of 6.7 ppm/°C at 25°C. Because of its refractoriness, it is difficult and costly to process by standard metallurgical practices. However, it is produced readily by CVD.

14.1 CVD Reactions

Rhenium is generally deposited by either the hydrogen reduction or by the thermal decomposition of its halides (45). The structure of the rhenium atom is such that it possesses all valence states from Re^{-1} to Re^{+7} of which Re^{+7} is the strongest and characteristic valence. This multivalence results in the formation of a wide series of halides and at least three fluorides and up to six chlorides have been indentified. However, with these halides, the maximum valence state of 7 is never realized, probably due to the fact that the rhenium atom is not capable of accommodating 7 atoms of the halogen. The hydrogen reduction of the fluoride is as follows:

[1] $ReF_6 + 3H_2 \longrightarrow Re + 6HF$

Deposition temperature range is 500-900°C with best deposits obtained at 700°C. At the low end of the temperature range, fluoride compounds remain incorporated in the deposit. Pressure is usually < 20 Torr. Equation [1] is a simplified version of a reaction that involves the formation of subfluorides such as ReF_4.

Grain refinement can be obtained by the addition of a small amount of H_2O in the gas stream (44).

The hydrogen reduction of the chloride is not normally used since it produces gas phase precipitation which is almost impossible to avoid.

Another common reaction is the pyrolysis of the chloride:

[2] $2ReCl_5 \longrightarrow 2Re + 5Cl_2$

Deposition temperature range is 1000-1250°C and best deposits are obtained at low pressure (< 20 Torr). This reaction usually gives a more ductile and purer material than does Reaction 1 although higher temperature is necessary. The chloride is usually prepared *in situ* by direct chlorination by heating the metal at 500-600°C (see Chapter Four, Section 3.1) (3).

Rhenium can also be deposited by the pyrolysis of the carbonyl:

[3] $Re_2(CO)_{10} \longrightarrow 2Re + 10CO$

Deposition temperature range is 400-600°C at a pressure of 200 Torr. Rhenium carbonyl is a solid at room temperature and must be vaporized at a temperature > 117°C. It decomposes at 250°C. This reaction is used for the coating of spheres in a fluidized bed (47).

14.2 Applications

- Heaters for high temperature furnaces

- Boats, crucibles, tubes and other free-standing shapes

- Contacts and diffusion-barrier metallization and selective deposition on silicon in semiconductor applications (experimental)

• Thermocouples

15.0 RHODIUM AND RUTHENIUM

Rhodium is a hard and durable metal with high reflectance, high resistance to corrosion and with low electrical resistivity (4.51 μohm-cm. at 20°C). It has a face centered cubic structure, a high melting point of 1966°C, a density of 12.41 g/cm^3, a thermal expansion of 8 ppm/°C at 25°C and a thermal conductivity of 1.50 W/cm.°C at 25°C.

15.1 CVD Reactions

The rhodium halides, like those of the other platinum group metals, are volatile with a decomposition point too close to the vaporization point to make them usable for CVD transport.

Rhodium is commonly produced by the decomposition of metallo-organic precursors such as rhodium acetyl acetonate, $Rh(CH_3COCHCOCH_3)_3$. This precursor evaporates at 230°C and the metal is deposited at 250°C and at 1 atm. pressure. Another precursor is rhodium trifluoro-acetyl acetonate, $Rh(C_5H_4F_3O_2)_3$. The reaction takes place at 400°C and at 1 atm. pressure (48).

Rhodium is also produced by the decomposition of the carbonyl, $Rh_4(CO)_{12}$, which decomposes above 95°C, the metal being deposited at approximately 250°C (49,50).

Note: Ruthenium is another platinum group metal with properties and CVD processing very similar to rhodium (51).

15.2 Applications

- Electrical contacts and diffusion barriers in IC's

16.0 TANTALUM

Tantalum is a refractory metal with a high melting point (2996°C) which is very chemically resistant especially to acid corrosion. It is ductile when pure. It has a body centered cubic structure and a density of 16.65 g/cm³ and is similar to niobium. It has a thermal expansion of 6.5 ppm/°C at 15°C, a thermal conductivity of 0.54 W/cm.°C at 25°C and an electrical resistivity of 12.45 μohms/cm at 25°C. It is produced extensively by CVD.

16.1 CVD Reactions

The direct decomposition of the halides is readily accomplished but requires high temperatures. For that reason their reduction with hydrogen is preferable (52,53). The chloride is generally used. It is generated in situ (see Chapter 4,Section 3.1) at 550°C as follows:

$$Ta + 2\ 1/2Cl_2 \longrightarrow TaCl_5$$

Deposition by hydrogen reduction occurs in the temperature range of 900-1300°C at low pressure (ca. 10 Torr). Tantalum has a high affinity for hydrogen and, for that reason, it is preferable to heat and cool the reactor in an inert atmosphere such as argon. The hydrogen reduction of the bromide or iodide is also feasible but is used less frequently.

16.2 Applications

- Thin film capacitors

- Corrosion resistant coatings

- Ordnance devices

17.0 TIN

The common tin or white β tin has a tetragonal structure, a density of 7.3 g/cm^3 and a low melting point of 232°C. It is generally chemically inert and oxidation resistant. It has a thermal expansion of 20 ppm/°C at 25°C, a thermal conductivity of 0.6 W/cm.°C at 25°C and an electrical resistivity of 11 μohm-cm at 0°C. It is deposited by CVD on an experimental basis, but because it must be deposited above its melting point, no practical applications are known at this time.

The hydrogen reduction of the chloride has been used to produce tin at 400-500°C as follows (3)(54):

$$SnCl_2 + H_2 \longrightarrow Sn + 2HCl$$

It is also deposited from the alkyls such as the tetramethyl or triethyl tin, $Sn(CH_3)_4$, $Sn_2(C_2H_5)_3$, at 500-600°C and usually low pressure (55).

18.0 TITANIUM

Titanium is a high strength metal with low density (4.54 g/cm^3) and excellent corrosion resistance. It has an hexagonal structure (alpha), which changes to cubic (beta) in a sluggish transformation at 880°C. Its melting point is 1660°C. It has a thermal expansion of 8.5 ppm/°C at 25°C, a thermal conductivity of 0.21 W/cm.°C at 25°C and an electrical resistivity of 42 μohm- cm at 20°C.

18.1 CVD Reactions

The thermal decomposition of the iodide or bromide is used to deposit titanium in the temperature range of 1200-1500°C as follows:

[1] $TiI_4 \longrightarrow Ti + 4I$

Note: Iodine is monatomic at these temperatures

The magnesium reduction is used extensively in the industrial production of the metal as follows:

[2] $TiCl_4 + 2Mg \longrightarrow Ti + 2MgCl_2$

This reaction cannot be considered as a controlled CVD reactions although some CVD mechanisms may be involved. A possible mechanism would be the reaction of $TiCl_4$ gas with solid or molten Mg, and then with Mg dissolved in liquid $MgCl_2$.

The hydrogen reduction of the chloride is used to deposit coatings. The chloride, $TiCl_4$, which is liquid at room temperature, can be transported in the reactor by a stream of hydrogen or prepared *in situ* by passing HCl, highly diluted with H_2, over titanium chips heated at 750-800°C as follows:

[3] $Ti + 4HCl \longrightarrow TiCl_4 + 2H_2$

Deposition of the titanium occurs mostly by the disproportionation of the lower chlorides in an essentially inert hydrogen although the hydrogen reduction of the chloride may also occur.

To prevent the absorption of the hydrogen by the titanium (potentially all the way to TiH_2 in thin films), an inert atmosphere such as argon is introduced before cooling.

Titanium can also be produced by the thermal decomposition of metallo-organics such as tris-(2.2'bipyridine) titanium at < 600°C. The deposit tends to retain impurities such as C, N_2 and H_2 (56).

18.2 Applications

- Production of metal foils and shapes

- Corrosion resistant coatings on steel and other substrates

- Preparation of titanium aluminides

19.0 TUNGSTEN

Tungsten is a refractory metal with a body-centered cubic (b.c.c.) structure, a very high melting point (3410°C), high density (19.3 g/cm³), low thermal expansion (4.45 ppm/°C at 25°C), high thermal conductivity (1.73 W/cm.K at 20°C) and low electrical resistivity (5.65 μohm-cm at 20°C). It is very resistant chemically but oxidizes readily. It is a brittle metal mostly because of impurities and is difficult to form by standard metallurgical processes. It can be produced easily by CVD as a very pure and relatively ductile metal used in many applications.

19.1 CVD Reactions

Tungsten is usually obtained by the hydrogen reduction of the halides as follows:

[1] $WF_6 + 3H_2 \longrightarrow W + 6HF$

This reaction has a ΔG = -33 Kcal/mol at 800K and an activation energy of 67 KJ/mol. The deposition temperature range is 300-700°C and the pressure range is 10 Torr-1 atm. Above 450°C, the reaction is diffusion controlled and relatively insensitive to temperature. Lower temperature (500°C) gives a finer grain structure with high strength (83 MPa) than high temperature (700°C) (57). Grain refinement is obtained with alternate deposition of silicon (see Chapter Two, section 4.3). This reaction is used extensively in semiconductor applications.

Another reaction is the silicon reduction of the fluoride:

[2] $2WF_6 + 3Si \longrightarrow 2W + 3SiF_4$

This reaction has a ΔG = -153 Kcal/mol W @ 800K. The deposition temperature range is 310-540°C and the pressure range 1-20 Torr. It is normally used on a silicon wafer substrate which becomes the source of silicon. It is self-limiting since it depends on the diffusion of WF_6 through the tungsten layer which can reach a maximum thickness of approximately one micron (58, 59). Reaction 2 has a much greater driving force than Reaction 1 and will always occur first.

The selective deposition of tungsten can be obtained with the following reactions (60):

[3] $WF_6 + 2SiH_4 \longrightarrow W + 2SiHF_3 + 3H_2$

[4] $WF_6 + 1\ 1/2\ SiH_4 \longrightarrow W + 1\ 1/2 SiF_4 + 3H_2$

In contract with Reaction 3, Reaction 4 is not strongly affected by the nature of substrate which can be tungsten or SiO_2.

Another deposition reaction is the hydrogen reduction of the chloride as follows:

[5] $WCl_6 + 3H_2 \longrightarrow W + 6HCl$

This reaction has a $\Delta G = -108$ Kcal/mol W at 800K. The deposition temperature range is 900-1300°C and the pressure range: 15-20 Torr. This reaction produces high purity deposits and is used to coat X-ray targets and produce structural parts. WCl_6 can be obtained by the direct chlorination *in situ* of the metal (see Chapter Four, section 3.1).

Tungsten can also be obtained by the pyrolysis of the carbonyl as follows:

[6] $W(CO)_6 \longrightarrow W + 6CO$

The deposition temperature range is 250-600°C and the pressure is up to 20 Torr. To minimize carbon inclusion, the carbonyl is highly diluted with hydrogen (ratio of 1/100) and high purity deposits can be obtained (3). Reaction 6 has a low temperature reaction and is being considered for deposition on silicon and III-V semiconductors.

In addition to the thermal CVD reactions listed above, tungsten can be deposited by plasma CVD using Reaction 1 at 350°C (61,62). At this temperature, a metastable alpha structure is formed instead of the stable b.c.c. (63). Tungsten is also deposited by an Excimer laser by reaction 1 at < 1 Torr to produce stripes on silicon substrate (64).

19.2 Applications

• Replacement of aluminum and general metallization of integrated circuits (IC's)

• Selective deposition, via plugs and gate electrodes for very large scale integrated circuits (VLSI)

• Diffusion barriers between silicon and aluminum in IC's

• Thermionic cathodes (co-deposited with thorium) (65)

• Coating for targets for X-ray cathodes (co-deposition with rhenium)

• Fabrication of free standing shapes (tubes, crucibles, plates) (66)

• Fabrication of furnace mufflers (67)

• Fabrication of free standing hollow spheres with thin wall ($<5 \mu$m) (68)

• Selective absorber coatings for solar energy collectors (69)

• Ordnance devices

20.0 INTERMETALLICS

Intermetallics are compounds of two metals in which a progressive change in composition is accompanied by a progression of phases differing in crystal structure. Many intermetallics have high strength at high temperature, high stiffness and low density.

Their limitations are generally low room temperature ductility and low fracture toughness. Some of the most important intermetallics are the aluminides of titanium, nickel and iron. Several intermetallics have been produced by CVD on an experimental basis as shown below.

20.1 Titanium Aluminides

There are two important titanium aluminides: Ti_3Al which has a hexagonal structure with a density of 4.20 g/cm^3 and a melting point of 1600°C and TiAl which has a tetragonal structure with a density of 3.91 g/cm^3 and a melting point of 1445°C. As do all aluminides, they have excellent high temperature oxidation resistance owing to the formation of a thin alumina layer on the surface. They have potential applications in aerospace structures (70).

Titanium aluminides are produced by CVD from the halides by the following sequence of reactions (71):

[1] $3HCl + Al \longrightarrow AlCl_3 + 1\frac{1}{2}H_2$

[2] $HCl + AlCl_3 + Ti \longrightarrow TiCl_3 + AlCl + \frac{1}{2}H_2$

[3] $3AlCl \longrightarrow AlCl_3 + 2Al$

[4] $1\frac{1}{2}AlCl + TiCl_3 \longrightarrow 1\frac{1}{2}TiCl_3 + Ti$

[5] $Ti + Al \longrightarrow TiAl$

[6] $3Ti + Al \longrightarrow Ti_3Al$

20.2 Ferro Nickel

Ferro nickel is produced by the co-decomposition of the carbonyls as follows (72):

$$xFe(CO)_5 + yNi(CO)_4 \longrightarrow Fe_xNi_y + (5x + 4y)CO$$

20.3 Nickel-Chromium

Nickel-chromium is produced by reaction of $CrCl_2$ with a nickel surface in a hydrogen atmosphere as follows (73):

$$xNi + yCrCl_2 + yH_2 \longrightarrow Ni_xCr_y + 2yHCl$$

20.4 Tungsten-Thorium

Tungsten-thorium intermetallics can be used as a long life thermionic cathode emitter for high power applications in high frequency tubes. They are obtained by CVD from tungsten fluoride and a thorium metallo-organic such as thorium hepta-fluorodimethyl octanedione, $Th(C_{10}H_{10}F_7O_2)_4$ in a codeposition reaction at 800°C and 10-100 Torr (65).

20.5 Niobium-Germanium

Niobium germanide, Nb_3Ge, is a superconductor with a high transition temperature (T_C= 20K). It is prepared by CVD by the co-reduction of the chlorides as follows:

$$3NbCl_5 + GeCl_4 + 9\frac{1}{2}H_2 \longrightarrow Nb_3Ge + 19HCl$$

The chlorides are prepared *in situ* and the deposition reaction

154   CHEMICAL VAPOR DEPOSITION

temperature is 900°C (74,75).

REFERENCES

1. Green, M.L. and Levy, R.A., CVD of W and Al for VLSI Applications, *Proc. 5th European Conf. on CVD,* (J. O. Carlsson and J. Lindstrom, Eds.), 441-447, Uppsala Univ., Sweden (1985)

2. Pierson, H.O, Aluminum Coatings by the Decomposition of Alkyls, *Thin Solid Films,* 45;257-264 (1977)

3. Powell, C.F., Chemically Deposited Metals, in *Vapor Deposition,* (C. F. Powell et al, Eds.), John Wiley & Sons, New York (1966)

4. Calloway, A.R., Galantowicz, T.A. and Fenner, W.R., Vacuum UV Driven CVD of Localized Aluminum Thin Films, *J. Vac. Sci. Technol.,* A 1 (2(1)):534-536 (Apr-June 1983)

5. Tsao, J.Y. and Erlich, D.J., Patterned Photonucleation of CVD of Aluminum by UV Laser Photodeposition, *Appl. Phys. Lett.,* 45(6):617-619 (15 Sept. 1984)

6. Motooka T., Gorbatkin, S., Lubben, D., Eres, D. and Greene, J.E., Mechanisms of Al Film Growth by UV Laser Photolysis of Trimethylaluminum, *J. Vac. Sci. Technol.* a4(6):3146-3152 (Nov/Dec 1986)

7. Oprysko, M.M. and Beranek, M.W., Nucleation Effects in Visible Laser CVD, *J. Vac. Sci. Technol.,* B(5)2:496-503 (Mar/Apr 1987)

8. Higachi, G.S. and Fleming, C.G., Patterned Aluminum Growth

via Excimer Laser Activated MOCVD, *App. Phys. Lett.,* 48(16):1051-1053 (23 Apr. 1986)

9. Kato, T., Ito, T., Ishikawa, H. and Maeda, M., Magnetron-Plasma CVD System and its Applications to Aluminum Film Deposition, *Proc. Int. Conf. on Solid State Devices and Materials,* Business Center for Academic Societies, Tokyo, Japan (Aug. 1986)

10. Wood, J.M. and Frey, F.W., CVD of Beryllium Metal, *Proc. of Conf. on CVD of Refractory Metals*, 205-216, AIME, NY (1967)

11. Bauerle, D., Laser Induced Chemical Vapor Deposition, in *Laser Processing and Diagnostics,* (D. Bauerle, Ed.), Springer-Verlag, New York (1984)

12.Mazille, H.M.J., CVD of Chromium onto Nickel, *Thin Solid Films,* 65:67-74 (1980)

13. Hanni, W. and Hintermann, H.E., CVD of Chromium, *Thin Solid Films*, 40:107-114 (1977)

14. Anantha, N.G., et al, Chromium Deposition from Dicumene Chromium to Form Metal Semiconductor Devices, *J. Electrochem. Soc.*, 118(1):163-165 (1971)

15. Erben, E., Bertinger, R., Muhlratzer, A., Tihanyl, B. and Cornils, B., CVD Black Chrome Coatings for High Temperature Photothermal Energy Conversion, *Sol. Energy Mater.,* 12(3):239-248 (July.Aug. 1985)

16. Pawlyk, P., Gas Plating Metal Objects with Copper ac.ac, *US Patent 2704728* (Mar 22 1955)

17. Van Hemert, R.L. et al, Vapor Deposition of Metals by Hydrogen Reduction of Metal Chelates, *J. Electrochem. Soc.,*

112(11):1123-1126 (1965)

18. Houle, F.A., Jones, C.R., Baum, T., Pico, C. and Kovac, C.A., Laser CVD of Copper, *Appl. Phys. Lett.*, 46(2): 204-206 (15 Jan. 1985)

19.Larson, C.E., Baum, T.H. and Jackson, R.L., Chemical Vapor Deposition of Gold, *J. Electrochem. Soc.*, 134(1): 266 (Jan. 1987)

20. Baum, T.H., Laser Chemical Vapor Deposition of Gold. The Effect of Organometallic Sctructure, *J. Electrochem. Soc.*, 134(10):2616-2619 (Oct. 1987)

21. Kodas, T.T., Baum, T.H. and Comita, P.B., Kinetics of Laser-Induced Chemical Vapor Deposition of Gold, *J. Appl. Phys.*, 62(1):281-286 (1 July 1987)

22. Macklin, B.A. and Withers, J.C., *Proc. of Conf. on CVD of Ref. Metals,* 161-173, Gatlinburg, TN, Amer. Nuclear Soc., Hinsdale, IL (1967)

23. Harding, J.T., Kazarof, J.M. and Appel, M.A., Iridium Coated Thrusters by CVD, NASA T.M.101309, in *Proc. of Second Int. Conf. on Surface Modification Technologies*, AIME and ASM, Chicago, IL, (Sept. 1988)

24. Smith, D.C., Pattillo, S., Elliot, N., Zocco, T., Laia, J. and Sattelberger, A., Low Temperature Organometallic Chemical Vapor Deposition (OMCVD) of Rhodium and Iridium Thin Films, *Proc. 11th Int. Conf. on CVD*, (K. E. Spear and G. W. Cullen, Eds.), 610-616, Electrochem. Soc., Pennington, NJ 08534 (1990)

25. Kaplan, R. and Bottka, N., Epitaxial Growth of Iron on GaAs by MOCVD in Ultra High Vacuum, *App. Phys. Lett.* 41(10):972-974 (1982)

26. Crawford, J., Refractory Metals Pace IC Complexity, *Semiconductor Int.*, 84-86 (March 1987)

27. Pauleau, Y., CVD of Refractory Metals and Refractory Metal Silicides, *Proc. 10th. Int. Conf. on CVD,* (G. Cullen, Ed.), 685-699, Electrochem. Soc., Pennington, NJ 08534 (1987)

28. Isobe, Y., Yamanaka, S., Son, P. and Miyake, N., Structure and Thermal Resistance of Chemically Vapor Deposited Molybdenum on Graphite, *J. Vac. Sci. Technol.*, A 4(6): 3046-3049 (Nov. Dec. 1986)

29. Isobe, Y., Shirakawa, H. Son, P. and Miyake, M., Thermal Shock Resistance of Chemically Vapor-deposited Molybdenum Coatings on Graphite, *J. of the Less Common Metals*, 152:251-260 (1989)

30. Kaplan, L.H. and d'Heurle, F.M., *J. Electrochem. Soc.* 117(5):693 (1970)

31. Ianno, N.J. and Plaster, J.A., Plasma Enhanced CVD of Molybdenum, *Thin Solid Films,* 147(2):193-202, (2 Mar. 1987)

32 .Sheward, J.A. and Young, W.J., The Deposition of Molybdenum and Tunsten Coatings on Gun Steel Substrates by a Plasma Assisted CVD Technique, *Vacuum*, 36(1-3):37- 41 (Jan. Mar. 1986)

33. Carver, G.E., CVD Molybdenum Thin Films in Photothermal Solar Converters, *Solar Energy Mat.*, 1(56):357-367 (June-Aug. 1979)

34. Carver, G.E. and Seraphim, B.O., CVD Molybdenum Thin Films for High Power-Laser Mirrors, *Laser Induced Damage in Optical Materials*, Boulder CO, Publ. of NBS, Wash. DC 20402 (Oct. 30-31, 1979)

35. Jenkins, W.C., *U.S. Patent 3355318,* Nov. 28, 1967

36. Owen, L.W., Observations on the Process of Nickel Deposition by Thermal Decomposition of Nickel Carbonyl at Low Pressure, *Metallurgia,* 165-173, (April 1959)

37. Van den Brekel, C.H.J., et al, CVD of Ni, TiN and TiC on Complex Shapes, *Proc. 8th Int. Conf on CVD* (J. M. Blocher Jr. et al, Eds.) 143-156, Electrochem. Soc., Pennington, NJ 08534 (1981)

38. Van Hemert, R.L., et al, Vapor Deposition of Metals by the Hydrogen Reduction of Metal Chelates, *J. Electrochem. Soc.,* 112(11):1123-1126 (1965)

39. Krauter, W., Bauerle, D. and Fimberger, F., Laser Induced CVD of Nickel by the Decomposition of $Ni(CO)_4$, *Appl. Phys.,* A31(1:13-18 (May 1983)

40. Jervis, T.R. and Newkirk, L.R., Metal Film Deposition by Laser Breakdown CVD, *J. Mater. Res.,* 1(3):420-424, (May-June 1986)

41. Skibo, M. and Greulich, F., Characterization of CVD Ni- (0.05-0.20wt%)B Alloys, *Thin Solid Films,* 113:225-234 (1984)

42. Miyake, M., Hirooka, Y., Imoto, R. and Sano, T., Chemical Vapor Deposition of Niobium on Graphite, *Thin Solid Films,* 63(2):303-308 (1 Nov. 1979)

43. Rand, M.J., Chemical Vapor Deposition of Thin-Film Platinum, *J. Electrochem. Soc.,* 120(5):686-693 (May 1973)

44. Bauerle, D., Laser Induced Chemical Vapor Deposition, in *Laser Processing and Diagnostics,* (D. Bauerle, Ed.), Springer-Verlag, New York (1984)

45. Sherwood, P.J., The Relation between the Mechanical Properties of CVD Tungsten and Rhenium, *Proc. 3d. Int. Conf. on CVD*, (F. Glaski, Ed.), pp. 728-739, Am. Nucl. Soc. (1972)

46. Donaldson, J.G., A Preliminary Study of the Vapor Deposition of Rhenium and Rhenium-Tungsten, *J. of the Less Common Metals*, 14:93-101 (1968)

47. McCreary, W.J., Microspherical Laser Target by CVD, *Proc. 5th. Int. Conf on CVD*, (J. M. Blocher Jr. and H. E. Hintermann, Eds.) pp. 714-725, Electrochem., Soc. Pennington, NJ 08534 (1975)

48. Bernard, C., Madar, R. and Pauleau, Y., Chemical Vapor Deposition of Refractory Metal Silicides for VLSI Metallization, *Solid State Tech.*, 79-84, (Feb. 1989)

49. Fillman, L.M. and Tang, S.C., Thermal Decomposition of Metal Carbonyls: a Thermogravimetry-Mass Spectrometry Study, *Thermochimica Acta*, 75:71-84 (1984)

50. Smith, D.C., Pattillo, S., Elliott, N., Zocco, T., Laia, J. and Sattelberger, A., Low Temperature Organometallic Chemical Vapor Deposition (OMCVD) of Rhodium and Iridium Thin Films, *Proc. 11th Int. Conf. on CVD*, (K. E. Spear and G. W. Cullen, Eds.) 610-616, Electrochem. Soc., Pennington NJ 08534 (1990)

51. Gross, M.E., Papa, L.E., Green, M.L. and Schnoes, K.J., Chemical Vapor Deposition of Ruthenium, *Extended Abstract Bull.*, No. 101, 165th Electrochem. Soc. Meeting, Cincinnati, OH (1984)

52. Perry, A.J., Beguin, C. and Hintermann, H.E., Tantalum Coating of Mild Steel at Atmospheric Pressure, *Thin Solid Films*, 66:197-210, (1980)

53. Bryant, W.A. and Meier, G.H., Preparation of Tantalum Based

Alloys by a Unique CVD Process, *Proc. 5th. Int. Conf on CVD*, (J. M. Blocher Jr. and H. E. Hintermann, Eds.) pp. 161-177, Electrochem. Soc., Pennington, NJ 08534 (1975)

54. Audosio, S., Chemical Vapor Deposition of Tin on Iron or Carburized Iron, *J. Electrochem. Soc.*, 127(10):2299-2304 (Oct. 1980)

55. Homer, H.J. and Cummins, O., Coating with Tin by Thermal Decomposition of Organotin Vapors, *US. Patent* 2916400 (Dec. 8,1959)

56. Morancho, R., Petit, J., Dabosi, F. and Constant, G., A Corrosion Resistant Titanium Rich Deposit Prepared by CVD at Low Temperature from Tris-(2.2'Bipyridine) Titanium, *Proc. 7th Int. Conf. on CVD*, (T. O. Sedgwick and H. Lydtin, Eds.), 593-603, Electrochem. Soc., Pennington, NJ 08534 (1979)

57. Green, M.L., The Current Statute of CVD Tungsten as an Integrated Circuit Metallization, *Proc. 10th. Int. Conf. on CVD*, (G. Cullen, Ed.), 603-611, Electrochemical Soc., Pennington, NJ 08534 (1987)

58. Tsao, K.Y. and Busta, H.H., Low Pressure CVD of Tungsten on Polycrystalline and Single-Crystal Silicon via the Silicon Reduction, *J. Electrochem. Soc.*, 131(11):2702-2708 (Nov. 1984)

59. Green, M.L., Ali.Y, Boone, T., Davidson, B., Feldman, L. and Nakahara, S., The Formation and Structure of CVD W Films Produced by the Si Reduction of WF_6, *J. Electrochem. Soc.*, 134(9):2285-2292 (Sept. 1987)

60. Kobayashi, N., Goto, H. and Suzuki, M., In situ IR Spectroscopy during Selective Tungsten Deposition, *Proc. 11th Int. Conf. on CVD* (K. E. Spear and G. W. Cullen, Eds.) 434-440 , Electrochem.

Soc., Pennington, NJ 08534 (1990)

61. Greenberg, K.E., Abnormal-glow-discharge Deposition of Tungsten, *App. Phys. Lett.*, 50(16):1050-1052 (20 April 1987)

62. Kember, P.N. and Astell-Burt, P.J., Plasma Enhanced CVD of Tungsten and Tungsten Silicide, *Vide, Couche Minces* 42-(236):167-173 (Mar-Apr. 1987)

63. Tang, C.C., Plasma Enhanced CVD of beta-Tungsten, a Metastable Phase, *App. Phys. Lett*, 45(6):633-635 (15 Sept. 1984)

64. Zhang, G.Q, Krypton Laser Induced CVD of Tungsten, *J. Appl. Phys.*, 62(2):673-675 (15 July 1987)

65. Gartner, G., Janiel, P. and Lydtin, H. CVD of Tungsten/Thorium Structures", *Proc. 5th. European Conf. on CVD*, (J. Carlsson and J. Lindstrom, Eds.), 51-60, U. of Uppsala, Sweden (1985)

66. Kaplan, R.and Tuffias, R., Large Seamless Tungsten Crucibles Made by CVD, *Int. J. Refract. Hard Metals*, 3(3) (15 Sept. 1984)

67. Swedberg, R.C., Buckman, R.W. and Bowen, W.W., W-Re Composite Tube Fabricated by CVD, *Thin Solid Films*, 72(2):393-398 (1 Oct. 1980)

68. Carroll, D.W. and McCreary, W.J., Fabrication of Thin-wall, Free Standing Inertial Confinement Fusion Targets by CVD, *J. Vac. Sci. Technol.*, 20(4):1087-1090 (Apr. 1982)

69. Erben, E., Muehlratzer, A., Tihanyi, B. and Cornils, B., Development of New Selective Absorber Coatings for High Temperature Applications, *Sol. Energy Mat.*, 9(3): 281-292 (Oct.-Dec. 1983)

70. Hunt, M., The Promise of Intermetallics, *Materials Engineering,* 35-39 (Mar. 1990)

71. Benander, R.F. and Holz, R.A., Deposition of Titanium Aluminide, *US Patent* No. 4698244, (Oct. 6, 1987)

72 Green, M.L. and Levy, R.A., Chemical Vapor Deposition of Metals for Integrated Circuit Aplications, *J. of Metals,* 62-71, (June 1985)

73. Hocking, M.G., Vasantasree, V. and Sidky, P.S., *Metallic and Ceramic Coatings,* Longman Scientific & Technical Ltd., London (1989)

74. Wan C.F. and Spear, K., CVD of Niobium Germanide from Partially Reacted Input Gases, *Calphad,* 7(2):149-155, (Apr.-June 1982)

75. Asano, T., Tanaka, Y. and Tachikawa, K., Effects of Deposition Parameters on the Synthesis of Nb_3Ge in the CVD Process, *Cryogenics,* 25(9):503-506 (Sept. 1985)

6

THE CVD OF NON-METALLIC ELEMENTS AND SEMICONDUCTORS

1.0 INTRODUCTION

This chapter is a review of the CVD of non-metallic elements, including boron and the allotropic forms of carbon, graphite and diamond. These elements are very important industrial materials and CVD is used extensively in their production. Also reviewed in this chapter is the CVD of semiconductors which includes the non metallic element silicon, the metalloid germanium and the III-V and II-VI compounds. The production of these materials by CVD is very extensive particularly in the semiconductor industry.

2.0 BORON

Boron is a light element with a density of 2.34 g/cm^3, an hexagonal structure, a high melting point (generally recognized as 2080°C), a thermal expansion of 8.3 ppm/°C and a thermal conductivity of 0.27 W/cm.°C at 20°C. It is an electrical insulator with a resistivity of 1.8 x 10^{12} μohm-cm at 20°C. It has good chemical resistance and is hard and brittle. It is produced by CVD in the form of coatings and fibers.

2.1 CVD Reactions

The hydrogen reduction of the chloride is the major production process (1,2) with a deposition reaction is as follows:

[1] $2BCl_3 + 3H_2 \longrightarrow 2B + 6HCl$

The deposition temperature range is 900-1300°C and the pressure is up to 1 atm.. The kinetics of this reaction were studied by Tanaka (3). This reaction is used to produce boron fibers on an industrial scale (see Chapter Thirteen).

The pyrolysis of the bromide has also been reported (4). The deposition reaction is as follows:

[2] $2BBr_3 \longrightarrow 2B + 3Br_2$

The deposition temperature range is 1000-1500°C and the pressure 1 Torr.

Boron is also obtained by the pyrolysis of diborane. Only moderate temperature is required (400-900°C) with a pressure up to 1 atm. (4,5):

[3] $B_2H_6 \longrightarrow 2B + 3H_2$

In addition to the thermal CVD reactions mentioned above, plasma CVD is used for the low temperature deposition of boron (6).

2.2 Applications

- Production of boron fibers on W or C core (see Chapter Twelve) (7)

- Coatings for the first wall of fusion reactor (8,9)

3.0 CARBON AND GRAPHITE

3.1 Structure of Carbon and Graphite

The carbon atom has four valence electrons and four vacancies in its outer shell (10). In the ground state, it has a $1s^2\ 2s^2\ 2p^2$ electron configuration. Such a configuration normally forms two covalent bonds with its two half-filled 2p orbitals. This explains why carbon has several allotropic forms which are graphite, microcrystalline carbon, diamond, lonsdaleite (a form detected in meteorites) and the recently discovered 60- atom carbon spherical molecule. These allotropes are sometimes found together. An example is diamond-like carbon (DLC) produced by low pressure synthesis which is actually a mixture of diamond and graphite or some form of microcrystalline carbon.

The structure of graphite is shown in Figure 1. The carbon atoms form continuous hexagons in stacked basal planes (ab directions). Within each basal plane, the carbon atom is strongly bonded to its three neighbors with a covalent bond having a bond strength of 524 KJ/mol. This atomic bonding is threefold coordinated and is known as sp^2. The hybridized fourth valence electron is bonded to an electron of the adjacent plane by a much weaker metallic-like bond of only 7 KJ/mol. The spacing between basal planes (3.35 Angstroms) is larger than the spacing between atoms in the plane (1.42 Angstroms). Such a configuration results in a very large anisotropy in the crystal which, in turn, results in anisotropy of the properties.

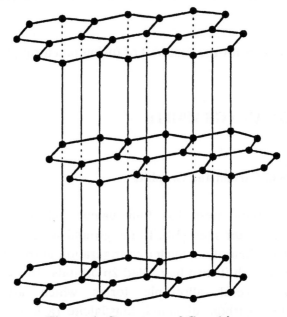

Figure 1. Structure of Graphite

The material obtained by CVD is usually an imperfect turbostratic graphite with warped basal planes and lattice defects. Its properties are generally similar to graphite but with even less bonding in the c direction. CVD graphite is also known as pyrolytic graphite or pyrographite.

3.2 Properties of CVD Graphite

Graphite has a very low thermal expansion in the ab plane but high in the c plane (1 ppm/°C vs. 15 ppm/°C in the 20-1000°C range). Likewise the thermal conductivity is high in the ab plane but very low in the c plane (1.90 W/cm.°C vs. 0.01 W/cm.°C at 25°C). The electrical resistance is low in the ab plane (5 μohm-cm at 20°C) but high in the c plane (3 ohm-cm). Graphite sublimes at 3600°C and retains its strength almost to that temperature. It is practically unaffected by thermal shock. It is very inert to most chemical environments except oxygen and it oxidizes readily above 500°C.

3.3 The CVD of Graphite

The CVD of graphite is relatively simple and is obtained by the thermal decomposition of a hydrocarbon. The most common precursor is methane, which is generally pyrolyzed at 1100°C in a pressure range of a few Torr to 1 atm. as follows (11):

$$CH_4 \longrightarrow C + 2H_2$$

Other common precursors are acetylene (C_2H_2) which decomposes in the 300-750°C temperature range and at pressures up to 1 atm. in the presence of a nickel catalyst. Another common precursor is propylene (C_3H_6), which decomposes in the 1000-1400°C temperature range at low pressure (100 Torr) (12).

The deposition of graphite can also be obtained by plasma CVD using a propylene-argon or a methane-argon mixture, in an RF plasma (0.5 MHz) at low pressure (10 Torr) and in the temperature range of 300-500°C (13).

3.4 Applications of CVD Graphite

- Boats and crucibles for liquid phase epitaxy

- Crucibles for molecular beam epitaxy

- Electrodes for plasma etching

- Reaction vessels for the gas phase epitaxy of III-V compounds

- Trays for silicon wafer handling

- Heating elements for high temperature furnaces

- Coating for fusion reactors

- Coating for nuclear fuel particles

- Chemical vapor infiltration of carbon-carbon structures (reentry heat shields, rocket nozzles and other aerospace components) (14)

- Aircraft disk brakes

- Coiled carbon fibers (15)

- Biomedical devices, heart valves, implants

4.0 DIAMOND AND DIAMOND-LIKE CARBON (DLC)

4.1 Characteristics and Properties of Diamond and DLC

Diamond is an allotropic form of carbon with a higher density than graphite (3.515 g/cm³) and a compact structure made of four-fold coordinated tetrahedral sp^3 bonds. This characteristic structure of diamond and the closeness of its atoms account for its outstanding properties; it is the hardest of all materials (ca. 10000 kg/mm²); it has the highest thermal conductivity (up to 18 W/cm.°C at 25°Cd). It has a low thermal expansion (2.8 ppm/°C at 0-400°C). It is the most perfectly transparent material and has one of the highest electrical resistivities (10^{16} ohm-cm at 25°C) and, when suitably doped, is an outstanding semiconductor material. It is very resistant to chemical attack but oxidizes rapidly above 600°C. These properties are characteristic of single crystal diamond.

Diamond can now be obtained as a polycrystalline material by CVD with properties similar to these of natural diamond. Efforts to produce single crystal thin films have so far been largely unsuccessful.

A special form of carbon, usually referred to as diamond-like carbon (DLC), is somewhat different with a structure that is essentially amorphous. DLC can be considered a hydrocarbon with a relatively high percentage of hydrogen that may reach 40 atomic %. It consists of a highly crosslinked network with isolated cluster dominated by trigonal coordinated sp^2 (graphite) carbon with some sp^3 (diamond). The ratio of sp^3 to sp^2 is a function of the hydrogen content. There is also some evidence of diagonal triple carbon bonding (sp hybridization). The material is usually referred to as a-C:H (16). The differences between diamond and DLC are summarized in Table 1.

The properties of DLC are generally similar to those of diamond such as high hardness and chemical inertness, but different in some key areas. As opposed to diamond, DLC has a variable index of refraction and variable electrical conductivity, both a function of hydrogen content. Like diamond, DLC can be obtained by CVD but its processing is less critical.

4.2 The CVD of Diamond

The chemical vapor deposition (CVD) of diamond is based on two factors which both require high energy: a) the carbon species must be activated since, at low pressure, graphite is thermodynamically stable and, without activation, only graphite would be formed, b) atomic hydrogen must be produced which selectively removes graphite and activates and stabilizes the diamond structure. Beside the need for atomic hydrogen, other factors such as energy input and the presence of oxygen have been shown to play an important role.

TABLE 1

Comparison of Diamond and DLC

	Diamond	DLC
Composition	Essentially pure carbon (H2 < 1%)	Up to 40 at.% hydrogen
Microstructure	Crystalline	Amorphous
Atom bonding state	sp3 tetrahedral	(sp2) graphite and (sp3) diamond
Raman spectroscopy	Sharp peak at 1332 cm	Broad humps at 1330 and 1550 cm

The deposition mechanism is complex and not fully understood at this time. The basic reaction involves the decomposition of a hydrocarbon such as methane as follows:

[1] $CH_4 \text{ (g)} \longrightarrow C \text{ (diamond)} + 2H_2 \text{ (g)}$
                         activation

This seemingly simple process is in actuality a complex occurrence where atomic hydrogen plays a crucial role (16). In the normal state, hydrogen is a molecule which dissociates in certain environments such as very high temperature (i.e. >2000°C) or in a high current density arc, to produce atomic hydrogen in a very endothermic reaction as follows:

[2] $H_2 \longrightarrow 2H$ $(\Delta H = 434.1 \text{ KJmol}^{-1})$

The rate of dissociation is a function of temperature, increasing rapidly above 2000°C. It also increases with decreasing pressure. The rate of recombination (i.e. the reconstitution of the molecule) is very rapid since the mean-free-path dependent half life of atomic hydrogen is only 0.3 sec.

Atomic hydrogen is extremely reactive (whereas molecular hydrogen is not). It etches graphite readily at a rate which is twenty times as high as the rate at which it etches diamond. The rate difference is even greater in the presence of oxygen. This is a very important characteristic since, when graphite and diamond are deposited together, graphite is preferentially removed while diamond remains.

Another important function of atomic hydrogen is its contribution to the stabilization of the sp^3 dangling bonds found on the diamond surface plane as it is formed. Without hydrogen, these bonds may not be maintained and the diamond {111} plane would collapse (flatten out) to the graphite structure (see Figure 2). These two characteristics, graphite removal and sp^3 bond stabilization, are essential factors in the growth mechanism of diamond.

Recent consensus among researchers is that diamond growth occurs mainly from primary species such as methyl radicals (CH_3) and acetylene (C_2H_2) (17). This has been confirmed experimentally on many occasions when diamond has been deposited from a large variety of precursors besides methane such as aliphatic and aromatic hydrocarbons, alcohols and ketones, solid polymers such as polyethylene, polypropylene and polystyrene, as well as fluorocarbons.

This behavior is consistent with the fact that all those compounds decompose to the stable species mentioned above, i.e. acetylene and methyl radicals. Deposition directly from acetylene (C_2H_2) appears to increase the crystallinity of the diamond deposit (18).

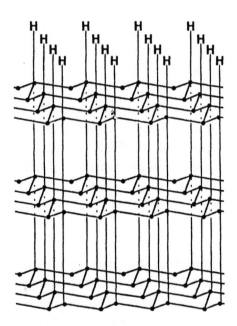

Figure 2. Structure of CVD Diamond Showing Hydrogen Dangling Bonds

The growth mechanism of diamond is best described by a model proposed recently by Frenklash and Spear which consists of two steps: an activation by hydrogen abstraction and an addition of acetylene as a monomer unit (19). In the first step, the diamond surface is activated by the removal of a surface bonded hydrogen ion by atomic hydrogen.

$$
\text{H.} \quad + \quad \overset{\displaystyle \text{C}}{\underset{\displaystyle \text{C}}{\overset{\backslash}{\underset{/}{\text{C}-\text{H}}}}} \longrightarrow \text{H}_2 \quad + \quad \overset{\displaystyle \text{C}}{\underset{\displaystyle \text{C}}{\overset{\backslash}{\underset{/}{\text{C.}}}}}
$$

The surface carbon radical is now activated. In the second step, it becomes the site for carbon addition by reacting with the carbon-hydrogen species in the gas phase.

$$
\overset{\displaystyle \text{C}}{\underset{\displaystyle \text{C}}{\overset{\backslash}{\underset{/}{\text{C.}}}}} \quad + \quad \text{C}_2\text{H}_2 \quad \longrightarrow \quad \overset{\displaystyle \text{C}}{\underset{\displaystyle \text{C}}{\overset{\backslash}{\underset{/}{\text{C}}}}}-\overset{\text{H}}{\underset{}{\text{C}}}=\overset{\text{H}}{\text{C.}}
$$

This model is consistent with many experimental observations appearing in the literature and should provide guidelines for the design of future experiments.

Indications are that the presence of oxygen or of an oxygen compound such as H_2O, CO, methanol, ethanol or acetone, is an important contributor to diamond film formation. The addition of a small amount of oxygen to methane and hydrogen tends to suppress the deposition of graphite by reducing the acetylene concentration. At the same time, the diamond growth rate is increased (20,21). In the case of water addition to hydrogen, the concentration of atomic hydrogen seems to increase, which explains the increased deposition rate and the suppression of graphite formation.

4.3 CVD Processes for Diamond

There are three major diamond CVD processes as follows:

Microwave plasma: A microwave plasma has an electron density typically 10^{20} electrons/m^3, and sufficient energy is developed in the plasma to dissociate hydrogen. A typical deposition apparatus is shown in Figure 3 (20). Gases are introduced at the top and flow past the substrate, which is generally a silicon wafer located in the lower part of the plasma. The substrate is sometimes heated by the interaction with the plasma and microwave power but more often separately by radiant or resistance heaters which allow more accurate control of the temperature. Typical reaction conditions are as follows:

Power:	600 W (incident)
Temperature of substrate:	800-1000°C
Gas mixture:	H_2/CH_4 50/1 to 200/1
Pressure:	0.1 to 45 Torr
Total gas flow:	20 to 200 cm³/m

With this deposition system, diamond is produced with a morphology and properties that vary as a function of the substrate temperature, the gas ratio, and the plasma intensity in the deposition zone. Deposition rate is low and averages one μm/hr. This low limit may be due to the restrictions on the amount of atomic hydrogen produced (estimated at 5%). Microwave deposition has the advantage of being very stable and can normally run for long periods of time (days) without interruption. However the plasma can be easily disturbed by the addition of oxygenated compounds.

Atomic hydrogen can also be generated with an RF plasma (13.56 MHz) but it seems that this process leads more to the deposition of diamond-like carbon (DLC).

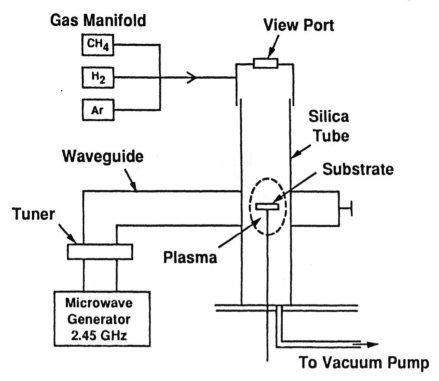

Figure 3. Schematic Diagram of Microwave Plasma CVD Apparatus for Diamond Deposition

Thermal Process (Hot Filament): A plasma can be generated by high temperature typically obtained by a heated tungsten filament, a tantalum wire or a tantalum tube (21). A schematic diagram of the equipment is shown in Figure 4. Atomic hydrogen is formed in contact with the hot metal which is kept at approximately 2000°C or slightly higher. The deposition rate and the composition and morphology of the deposit are a function of the temperature and the distance between the hot metal and the substrate. This distance is usually 1 cm or less. Much beyond that, most of the atomic hydrogen recombines and no diamond is formed.

Gas Manifold

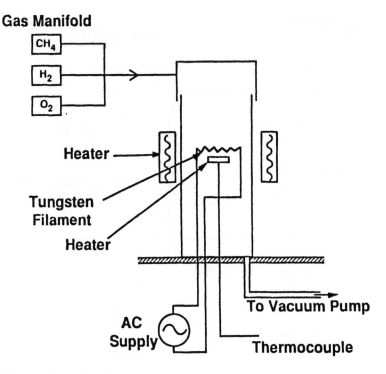

Figure 4. Schematic Diagram for Diamond Deposition by Thermal Process (Hot Filament)

The substrate is heated between 800 and 1000°C, and gas composition and other deposition parameters are similar to those used in microwave plasma deposition. Deposition rate is similarly reported as 0.5 to 1 μm/hour. A disadvantage of this method is the short life of the metallic heater, which tends to carburize, distort and embrittle, and can be used for only a short period of time. The heater also has a tendency to evaporate and contaminate the deposited diamond film. In addition, if an oxygen compound is added as mentioned above, tungsten (or any other refractory metal) cannot be used as it would oxidize readily. However, the equipment is of low cost and experiments are readily carried out. Other heating elements such as graphite or rhenium are being investigated.

Plasma arc deposition. In plasma arc deposition, a high intensity arc is generated between two electrodes by either DC, AC, microwave or high frequency current. A typical DC plasma deposition apparatus is shown in Figure 5 (22). Electrodes usually consist of a water cooled copper anode and a tungsten cathode. In the plasma discharge, temperature may reach 5000°C or higher. Several jet nozzles can be used simultaneously, and many design variations are possible including separate input nozzles for hydrogen and methane (the latter mixed with argon) and the feeding of these gases in a coaxial feed electrode. The plasma jet can be cooled rapidly just prior to coming in contact with the substrate by using a blast of cold inert gas fed into an annular fixture. Gaseous boron or phosphorus compounds can be introduced into the gas feed for the deposition of doped semiconductor diamond.

The very high temperatures obtained in an arc discharge allow an almost complete dissociation of the hydrogen molecules. As mentioned previously, the availability of hydrogen atoms is a key element in the formation of diamond. From this standpoint, arc discharge systems have an advantage over the other processes which produce a far smaller ratio of hydrogen atoms.

The formation of a high speed (supersonic) arc jet, which is the result of the sudden expansion of the gas as they are heated in the arc plasma, means that the atomic hydrogen and the carbon species and radicals are transported almost instantly to the deposition surface and the chances of hydrogen recombination and vapor phase reactions are minimized.

The abundant supply of atomic hydrogen may account for the extremely high deposition rate, in excess of 200 μm/hour, claimed for the arc discharge process, which is one to two orders of magnitude greater than the rate obtained by other deposition processes. Thick deposits have been produced by this process but their quality and purity is yet to be determined (23).

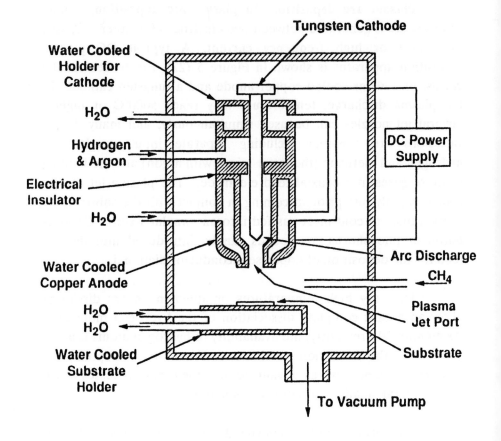

Figure 5. Schematic Diagram of Apparatus for DC Arc Discharge Unit for CVD of Diamond

One process for synthesizing diamond uses methane in which the carbon atom is a carbon 12 isotope enriched to 99.97% C^{12}. Diamond is deposited by CVD in a microwave plasma; the polyscrystalline deposit is crushed, processed to a single crystal with molten iron and aluminum at high temperature and pressure and recovered by leaching out the metals. The resulting C^{12} diamond is reported to have a 50% higher thermal conductivity than natural diamond (24).

4.4 The CVD of DLC

The deposition process for DLC differs from that of diamond in as much as the activation is not so much chemical (i.e. the use of hydrogen atoms) but physical. This physical activation is usually obtained by colliding accelerated ions produced by high frequency discharge which does not require and does not produce high temperature. Consequently the substrate remains relatively cool ($<300°C$) and a wide variety of substrate materials can be used including plastics.

This of course is a major advantage which the processing of diamond films does not possess, at least for the time being. The DLC deposit is amorphous with a sizable content of hydrogen and properties different from those of diamond coatings.

The deposition principle is relatively simple and a schematic diagram of the equipment is shown in Figure 6 (25). A high frequency RF gas discharge (13.56 MHz) is generated in a mixture of hydrogen and a hydrocarbon, such as methane or n-butane (C_4H_{10}). The asymmetry of the electrodes and the large difference in mobilities between electrons and ions result in the spontaneous generation of a negative potential on the substrate which, as a result, is bombarded by the ionized gas species.

The coating thus obtained can be a DLC material or a soft polymer- like graphite depending on the applied energy (25). The coating of a large number of substrates is possible as long as they can be electrically contacted. Deposition rate is 0.5 to 2 μm/hr.

Good surface preparation is essential as in all deposition processes and this can be achieved by chemically cleaning the substrate followed by sputter cleaning with argon just prior to the actual deposition.

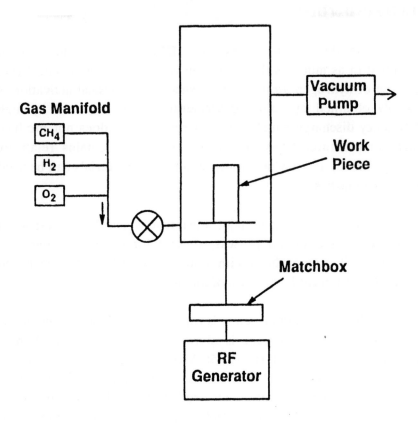

Figure 6. Schematic Diagram of DLC Deposition Apparatus

A process similar to the one just described has been developed at NASA Lewis Research Center (26). It uses a 30 cm. hollow cathode ion source with its optics masked to 10 cm. Argon is introduced to establish the discharge followed by methane in a 28/100 ratio of methane molecules to argon atoms. The energy level is 100 eV, the acceleration voltage 600 V and the resulting deposition rate 0.5 to 0.6 μm/hour.

With present deposition technologies, it is difficult, if not impossible, to produce thick DLC films, as they tend to delaminate and separate from the substrate when the thickness is greater than approximately 0.5 μm. This is the result of high internal compressive stresses which appear to be related to the hydrogen content of the material (27).

The deposition of DLC materials has reached the practical stage and is proceeding rapidly toward large industrial operations. Many applications are in the development stage and a number have reached a successful commercial stage.

4.5 Applications of Diamond and DLC

- Hard coatings for wear and abrasion resistance

- Coatings for cutting tools

- Heat sinks for electronics and opto-electronics

- Optical coatings

- Coatings for infra-red (IR) windows

- Electronic devices for severe environments (space, nuclear reactors, combustion engines, etc.)

- Sensors for severe environments

- Microwave devices

- Light-emitting diodes (LED)

- High speed electronic devices

5.0 SILICON

5.1 Properties

Silicon is a semiconductor material which is the basic building block of the present electronic industry. It has the crystal structure of diamond and its properties are influenced by the crystal orientation (28). CVD silicon can be either single crystal (epitaxial silicon), polycrystalline (polysilicon) or amorphous. Silicon has a density of 2.33 g/cm³, a melting point of 1410°C, a thermal expansion of 3 ppm/°C at 25°C and a thermal conductivity of 1.49 W/cm.°C at 25°C. It has an electrical resistivity of 3×10^6 μohm-cm. at 20°C and a bandgap of 1.1 eV. Semiconductor properties are imparted by doping its structure with boron, phosphorus or arsenic atoms. Silicon is relatively inert chemically but is attacked by halogens and dilute alkalies. It has good optical transmission especially in the infra-red.

5.2 CVD Reactions

There are a number of precursors and CVD reactions that can be used to deposit silicon. The deposit can be either single crystal (epitaxial), polycrystalline or amorphous.

Silicon Epitaxy. Silicon epitaxial films have superior properties. The applications are however limited by the high temperature of deposition which is generally above 1000°C. These reactions use chlorinated compounds of silicon (silicon tetrachloride, trichlorosilane or dichlorosilane) as precursors as follows:

[1] $SiCl_4 + 2H_2 \longrightarrow Si + 4HCl$

(on silicon wafer substrate, temperature range: 1150-1200°C)

[2] $SiCl_3H + H_2 \longrightarrow Si + 3HCl$

(on silicon wafer substrate, temperature range: 1080-1130°C)

[3] $SiCl_2H_2 \longrightarrow Si + 2HCl$

(in hydrogen atmosphere, on silicon wafer substrate, temperature range: 1030-1070°C)

Reactions 1, 2 and 3 are generally carried out at atmospheric pressure and produce films up to 100 μm thick. These reactions, which are used extensively in production, are reversible since the formation of HCl promotes the etching off of impurities during deposition due to the high energy states of silicon bonding at the sites of impurities (30).

As the density of devices crowded onto the silicon wafer increases, the problems of autodoping and interdiffusion become more acute and the high temperature limitation of the above reactions has prompted much experimental effort to develop epitaxial deposition systems that operate at lower temperature. This has been accomplished in the following experimental developments:

• deposition with silane (SiH_4) at very low pressure (1 to 15 mTorr) at 750-800°C and by plasma deposition with a temperature range of 600-800°C (30, 31)

• deposition with disilane (Si_2H_6) in an RF induction heated reactor at 850°C (32)

● deposition by rapid thermal heating with tungsten halogen lamp banks at 600-900°C (33)

● deposition by CO_2 laser at 700°C (34)

Polycrystalline Silicon (Polysilicon): Polycrystalline silicon is used extensively in semiconductor devices. It is normally produced by decomposition of silane at low pressure (ca. 1 Torr) as follows:

[4] $SiH_4 \longrightarrow Si + 2H_2$

This reaction is irreversible. At the temperature range of 1000-1040°C, epitaxial silicon is deposited but the deposit is generally unsatisfactory and the reaction is no longer used for that purpose. Polycrystalline silicon is obtained in the range of 610-630°C, which is close to the crystalline-amorphous transition temperature (35,36).

Amorphous Silicon: Amorphous silicon is generally deposited by Reaction 4 at a deposition temperature of 560°C and at low pressure (ca. 1 Torr) (34). Helium RF plasma CVD is also commonly used especially in the production of solar photovoltaic devices (37).

Production of Electronic Grade Silicon (EGS): As the first step in the production of electronic grade silicon (EGS), an impure grade of silicon is pulverized and reacted with anhydrous hydrochloric acid, to yield primarily tricholorosilane, $HSiCl_3$. This reaction is carried out in a fluidized bed at approximately 300°C in the presence of a catalyst. At the same time, the impurities in the starter silicon react to form their respective chlorides. These chlorides are liquid at room temperature with the exception of vanadium dichloride and iron dichloride, which are soluble in

$HSiCl_3$ at the low concentration prevailing. Purification is accomplished by fractional distillation.

The next step is the hydrogen reduction of the trichlorosilane (Reaction 2 above). The end product is a polycrystalline silicon rod up to 200 mm in diameter and several meters in length. The resulting EGS material is extremely pure with less than 2 ppm of carbon and only a few ppb of boron and residual donors. The Czochralski pulling technique is used to prepare large single crystals of silicon which are subsequently sliced into wafers for use in electronic devices (28).

5.3 Applications of CVD Silicon

Many applications of epitaxial silicon are found in IC's and other semiconductor devices such as digital bipolar, linear digital metal-on-silicon (MOS), discrete linear digital MOS and complimentary MOS (CMOS).

Polysilicon applications in integrated circuits are gate electrodes, interconnection conductors, resistor and emitter contacts. Other applications include thermal and mechanical sensors, and photovoltaic cells.

Amorphous silicon is mostly used in photovoltaic devices and photocopier drums.

6.0 GERMANIUM

6.1 Properties

Germanium was the semiconductor material used in the development of the transistor in the early 1950's. However, it exhibited high junction leakage current due to its narrow bandgap (0.66 eV) and is now largely replaced by silicon. Other detrimental factors are its low maximum operating temperature (100°C vs. 150°C for silicon) and the instability of its native oxide which is water soluble (28).

Germanium is a brittle metalloid element with a diamond structure, a density of 5.32 g/cm³, a melting point of 937°C, a thermal conductivity of 0.602 W/cm.°C at 25°C and an electrical resistivity of 4.6×10^7 μohm/cm at 22°C.

6.2 CVD Reactions

Single crystal germanium can be deposited by the hydrogen reduction of the chloride at 600-900°C as follows (38):

[1] $GeCl_4 + 2H_2 \longrightarrow Ge + 4HCl$

However, it is now produced mostly from the decomposition of germane, usually at atmospheric pressure and at temperatures ranging from 600 to 900°C as follows:

[2] $GeH_4 \longrightarrow Ge + 2H_2$

Crystalline deposits are obtained with Reaction 2 in the temperature range of 350-400°C at low pressure (< 1 Torr) (39). At still lower temperature (< 330°C) and moderate pressure (20-50 Torr), an amorphous germanium deposit is obtained (40). With Reaction (2), germanium is obtained by plasma CVD at 450°C and by laser CVD at 340°C (41,42).

Germanium is deposited from metallo-organic precursors such as tetrapropyl germanium, $Ge(C_3H_7)_4$ and tetraallyl germanium, $Ge(C_3H_5)_4$, with helium or hydrogen diluent. These reactions occur at low pressure and in the temperature range 575-700°C (43).

6.3 Applications

- Ge films on Si to tailor bandgap of heterostructures (39)

- Photovoltaic conversion (43)

- Alloys with silicon (44)

- Photodetectors

7.0 THE CVD OF III-V AND II-VI COMPOUNDS

III-V and II-VI compounds combine the elements of Group III (aluminum, gallium, indium) with those of Group V (phosphorus, arsenic, antimony, bismuth) and the elements of Group II (zinc, cadmium, mercury) with those of Group VI (sulfur, selenium, tellurium).

As can be imagined, a wide variety of compounds can be produced which, in addition to the binary materials, also include ternary and even quaternary materials.

These materials are very useful semiconductors and have a wide range of industrial applications, particularly in opto-electronics. One of their attractive features is the possibility of tailoring the band gap and the lattice constant in the ternary alloys by varying the composition (see Chapter Eight, section 5). CVD is now their major production process.

7.1 The III-V Compounds

The III-V compounds now being produced by CVD include: GaAs, GaN, GaP, InAs, InP, AlAs, BP, InGaAs, AlInAs, AlGaAs, GaPAs, InGaN and InGaPAs. The most common is gallium arsenide, GaAs, which has a cubic close packed structure (zinc blende) similar to that of silicon. The electronic band structure of GaAs gives it high electron mobility ($8000 \, cm^2/V$-sec. vs. $1500 \, cm^2/V$-sec. for Si). Unlike silicon, it can emit coherent light. It is very resistant to radiation (45).

The CVD of the III-V compounds is usually obtained by reacting an alkyl of a group-III element with a hydride of a group-V element. These reactions have largely replaced the co- reduction of the halides (46). The general reaction is as follows:

[1] R_nM + $XH_n \longrightarrow$ MX   +   nRH

    (group III)   (group V)   (III-V)   (hydrocarbon)

A specific example is:

[2] $(CH_3)_3Ga + AsH_3 \longrightarrow GaAs + 3CH_4$

These reactions usually occur at atmospheric pressure and in the following temperature range: 650-730°C for GaAs, 650-700°C for GaP, 630-750°C for InGaAs and 700-800°C for AlGaAs. The molar ratio of the III compound to the V compound is typically 1/10 (47). To obtain the desired semiconductor properties, dopants are added such as zinc (from diethyl zinc) or magnesium (from bis(cyclopentadienyl) magnesium) for p doping, and silicon (from silane) or selenium (from hydrogen selenide) for n doping.

Arsine (AsH_3) and phosphine (PH_3) are extremely toxic, so less hazardous substitutes such as tertiary butyl arsine, $C_4H_{11}As$, and tertiary butyl phosphine, $C_4H_{11}P$, are being considered (48).

Plasma CVD and thermal laser CVD are also used particularly in the deposition of GaAs. The formation of epitaxial GaAs at 500°C and polycrystalline GaAs at 185°C has been reported (49).

The reaction of the hydrides of both the Groups III and V elements is also used notably in the production of boron phosphide at 950-1000°C as follows (50):

[3] $B_2H_6 + 2PH_3 \longrightarrow 2BP + 6H_2$

7.2 The II-VI Compounds

The II-VI compounds now being produced by CVD include: ZnSe, ZnS, ZnTe, CdS, HgTe, CdMnTe and HgCdTe. Typical deposition systems include the reaction of a vaporized metal of group II (Zn, Cd, Hg, which all have low vaporization temperature), with an alkyl of the group VI element at a temperature range of 325- 350°C, as shown in the following examples:

[1] $(C_2H_5)_2Te + Cd (g) + H_2 \longrightarrow CdTe + 2C_2H_6$

[2] $(CH_3)_2Te + Hg (g) + H_2 \longrightarrow HgTe + 2CH_4$

Other II-VI compounds are deposited by the reaction of the Group VI hydride with the Group II vaporized metal (51):

[3] $H_2S + Zn (g) \longrightarrow ZnS + H_2$

[4] $H_2Se + Zn (g) \longrightarrow ZnSe + H_2$

Both reactions occur at 600-800°C and at pressures <100 Torr.

An excimer laser is used to deposit CdTe at low temperature (200°C) at 10-80 Torr by the following reaction (52):

[5] $(C_2H_5)_2Te + (CH_3)_2Cd \longrightarrow CdTe + hydrocarbons$

The properties and CVD of some of these compounds are also reviewed in Chapter Seven, Section 7.7.

7.3 Applications of III-V and II-VI Compounds

• Microwave devices

• Photo-chemical cells (53)

• Photovoltaic devices using CdTe (54)

• Light emitting diodes (LED)

• Solid state neutron detector ofboron phosphide which is a refractory semiconductor with a wide band gap (50)

• Field effect transistors (FET) of epitaxial InP

• Heterostructure bipolar transistors (HBT) of InGaAs and InAlAs

• IR transparent windows of ZnS, ZnSe and ZnTe

• Photoconductors of CdSe and CdS

These and other applications are reviewed in detail in Chapters Eight and Nine.

REFERENCES

1. Vandenbulcke, L. and Vuillard, G., CVD of Amorphous Boron on Massive Substrates, *J. Electrochem. Soc.*, 123(2):278-285 (1976)

2. Naslain, R., Thebault, J., Hagenmuller, P. and Bernard, C., The Thermodynamic Approach to Boron CVD Based on the Minimization of the Total Gibbs Free Energy, *J. Less Common Metals*, 67 (1), 85-100, (1979).

3. Tanaka, H., Nakanishi, N. and Kato, E., Kinetics of CVD of Boron on Molybdenum, *Proc. 10th. Int. Conf on CVD*, (G. Cullen, Ed.), Electrochem. Soc., Pennington, NJ 08534 (1987)

4. Combescure, C., Armas, B., Alnot, M. and Weber, B., Kinetic Study of Boron Tribromide Pyrolysis at Low Pressure, *Proc. 7th. Int. Conf. on CVD*, (T. Sedgwwick and H. Lydtin, Eds.) 351-359, Electrochemical Soc., Pennington, NJ 08534 (1979)

5. Pierson, H.O. and Mullendore, A.W., The CVD of Boron at Low Temperatures, *Thin Solid Films*, 83:87-91 (1981) 87-91

6. Chatterjee A. and White, D. Jr., Analysis of Plasma Enhanced CVD of Boron, *Proc. 5th European Conf. on CVD*, (J. Carlsson and J. Lindstrom, Eds.) 29-36, U. of Uppsala, Sweden (1985)

7. Carlsson, J. and Lundstrom, T., Mechanical Properties and Surface Defects of Boron Fibers Prepared in a Closed CVD System, *J. Mater. Sci.* 14(4):966-974 (1979)

8. Groner, P., Gimzewski, J. and Veprek, S., Boron and Doped Boron First Wall Coatings by Plasma CVD, *J. Nucl. Mater.*, 103(1-3):257-260 (1981)

9. Pierson, H.O. and Mullendore, R.W., Boron Coatings on Graphite for Fusion Reactor Applications, *Thin Solid Films*, 63:257-261 (1979)

10. Cotton, F. and Wilkinson, G., *Advanced Inorganic Chemistry*, Interscience Publishers, New York (1972)

11. Campbell, I. and Sherwood, E.M., *High Temperature Materials and Technology*, John Wiley & Son, New York (1967)

12. Gower, R.P. and Hill, J., Mechanism of Pyrocarbon Formation from Propylene, *Proc. 5th Int. Conf. on CVD*, (J. Blocher and H. Hintermann, Eds.), 114-129, Electrochem. Soc., Pennington, NJ 08534 (1975)

13. Inspektor, A., *J. Vac. Sci. Techn.*, A 4:375 (1986)

14. Buckley, J.D., Carbon-carbon, an Overview, *Ceram. Bul.*, 67-2:364-368 (1988)

15. Motojima, S., Kawaguchi, M., Nozaki, K. and Iwanaga, H., Preparation of Coiled Ceramic Fibers by CVD, *Proc. 11th. Int. Conf. on CVD* (K. Spear and G. Cullen, Eds.) 573-579, Electrochem. Soc., Pennington NJ 08534 (1990)

16. Deryagin, B. and Fedosev, D., Growth of Diamond and Graphite from the Vapor Phase, *Izd. Nauka*, Moscow, USSR (1977)

17.Spear, K., Diamond: Ceramic Coating of the Future, *J. Amer. Ceram. Soc.* 7-(2):171-91 (1989)

18. Yasuda, T., Ihara, M., Miyamoto, K., Genchi Y. and Komiyama, H., Gas Phase Chemistry Determining the Quality of CVD Diamond, *Proc. 11th. Int. Conf. on CVD*, (K. Spear and G. Cullen, Eds.) 134-140 Electrochem. Soc., Pennington NJ 08534 (1990)

19. Frenklash, M. and Spear, K., Growth Mechanism of Vapor Deposited Diamond, *J. of Mat. Res.*, 3(1):133-140 (1988)

20. Saito, Y., Sato, K., Tanaka, H., Fujita, K. and Matsuda, S., Diamond Synthesis from Methane, Hydrogen and Water Mixed Gas using a Microwave Plasma, *J. Mat. Sci.*, 223(3):842-46 (1988)

21. Yarborough, W., Diamond and DLC by CVD, *Proc. Conf. on High Performance Inorganic Thin Film Coatings,* GAMI, Gorham ME 04038 (1988)

22. Kuriha, K., et al, *App. Physics Letters,* 52(6) (1988)

23. Matsumoto, S., Deposition of Diamond from Thermal Plasma, *Proc. of the Materials Res. Soc.*, Spring Meeting, Reno, Nev. (April 1988)

24. Technology Update, *Ceramic Bull.*, 69-No.10, p.1639 (1990)

25. *ADLC Technical Brochure*, Bernex, Olten, Switz. (1988)

26. Mirtich, M., Swec, D. and Angus, J., Dual Ion Beam Deposition of Carbon Films with Diamond Like Properties, *Thin Solid Films*, 131:248-254 (1985)

27. Angus, J.C.,et al, Diamond and Diamond-like Phases Grown at Low Pressure; Growth, Properties and Optical Applications, *Diamond Optics*, SPIE 969:2-13 (1988)

28. Pearce, C.W., Crystal Growth and Wafer Preparation, in *VLSI Technology,* (S. M. Sze, Ed.), McGraw Hill Book Co., New York (1983)

29. Taylor, P.A., Silicon Source Gases for Chemical Vapor Deposition, *Solid State Technology*, 143-138 (May 1989)

30. Comfort, J.H. and Reif, R., Chemical Vapor Deposition of Epitaxial Silicon from Silane at Low Temperatures, *J. Electrochemn. Soc.*, 136-8:2386-2405 (Aug. 1989)

31. Reif, R., Low Temperature Silicon Epitaxy by Plasma Enhanced CVD, *Proc. 5th European Conf. on CVD,* (J. Carlsson and J. Lindstrom, Eds.) 13-19, Univ. of Uppsala, Sweden (1985)

32. Tsukune, A., Miyata, H., Mieno, F. and Furumura, Y., *Proc. 11th. Int. Conf. on CVD*, (K. Spear and G. Cullen, Eds.) 261-269, Electrochem. Soc., Pennington NJ 08534 (1990)

33. Green, M.L., Brasen, D. and Luftman, H., High Quality Homoepitaxial Silicon Films Deposited by Rapid Thermal Chemical Vapor Deposition, *J. Appl. Phys.*, 65(6)-15: 2558-2560 (1989)

34. Itoh, T. and Susuki, S., Interface Phenomena of Si on Si Epitaxy by CO_2 Laser CVD, *Proc. 10th. Int. Conf on CVD*, (G. Cullen, Ed.) 2124-223, Electrochem. Soc., Pennington, NJ 08534 (1987)

35.Yoon , H., Park, C. and Park, S., Structure and Electrical Resistivity of Low Pressure Chemical Vapor Deposited Silicon, *J. Vac. Sci. Technol.*, A4(6):3095 (1986)

36. Mulder, J., Eppenga, P., Hendriks, M. and Tong, J., An Industrial LPCVD Process for In Situ Phosphorus-Doped Polysilicon, *J. Electrochem. Soc.*, 137-1:273-279 (1990)

37. Ichikawa, Y., Sakai, H. and Uchida, Y., Plasma CVD of Amorphous Si Alloys for Photovoltaics, *Proc. 10th Int. Conf. on CVD* (G.Cullen, Ed.) 967-975, Electrochem. Soc., Pennington, NJ 08534 (1987)

38. Cave, E.F. and Czorny, B.R., *RCA Review,* (24):523, (1963)

39. Murota, J., Kobayashi, A., Kato, M., Mikoshiba, N. and Ono, S., Germanium Epitaxial Growth Mechanism in Low- Pressure CVD using Germane GAs, *Proc. 11th. Conf. on CVD*, (K. Spear and G. Cullen, Eds.), 325-331, Electrochem. Soc., Pennington, NJ 08534 (1990)

40. Allred, D. and Piontkowski, J., CVD Amorphous Germanium, Preparation and Properties, *Proc. 9th, Int. Conf on CVD*, (M. Robinson et al, Eds.) 546-557, Electrochem Soc., Pennington, NJ 08534 (1984)

41. Kember, P. and Moss, J., Deposition and Characteristics of Plasma Deposited Doped Germanium Films, *Proc. 10th Int. Conf. on CVD*, (G.Cullen, Ed.) 894-903, Electrochem. Soc., Pennington, NJ 08534 (1987)

42. Gow, T.R., Coronell, D.G. and Masel, R.I., The Mechanism of Laser-assisted CVD of Germanium, *J. Mater. Res.*, 4-3:634-640 (May/June 1989)

43. Morancho, R., Constant, G., Gallon, C., Boucham, J., Mazerolles, P. and Bernard, C., OMCVD Elaborating and Optical Properties of Germanium Carbon Alloys, *Proc. 5th European Conf. on CVD*, (J. Carlsson and J. Lindstrom, Eds.) 526-532, Univ. of Uppsala, Sweden (1985)

44. Kato, M., Cheng, M., Iwasaki, C., Murota, J., Mikoshiba, N. and Ono, S., Low Temperature Selective Silicon and Silicon-Germanium Epitaxial Growth in Ultraclean Low-pressure CVD, *Proc. 11th. Conf. on CVD*, (K. Spear and G. Cullen, Eds.) 240-246, Electrochem. Soc., Pennington, NJ 0853 (1990)

45. Vander Veen, M.R., Gallium Arsenide Sandwich Lasers, *Advanced Mat. & Processes*, 29-45 (1988)

46. Gandhi, S. and Bhat, I., Organometallic Vapor Phase Epitaxy: Features, Problems New Approaches, *MRS Bulletin,* 39-43, (Nov. 1988)

47. Lewis, C.R., MOCVD Adds Consitency to II-V Semiconductor Thin-layer Growth, *Res. & Dev.,* 106-110 (Nov. 1985)

48. Miller, G.A., Arsine and Phosphine Replacements for Semiconductor Processing, *Solid State Technology,* 59-60 (Aug. 1989)

49. Huelsman, A.D. and Reif, R., Plasma Deposition of GaAs Epitaxial Films from Metal-Organic Sources, *Proc. 10th Int. Conf. on CVD,* (G. Cullen, Ed.) 792-802, Electrochem. Soc., Pennington, NJ 08534 (1987)

50. Kumashiro, Y., Oikada, Y., Misawa, S. and Koshiro, T., The Preparation of BP Single Crystals, *Proc. 10th Int. Conf. on CVD,* (G. Cullen, Ed.) 813-817, Electrochem. Soc., Pennington, NJ 08534 (1987)

51. Collins, A. and Taylor, R., Optical Characterization of Polycrystalline SnS Produced via Chemical Vapor Deposition, *Proc. 11th. Conf. on CVD,* (K. Spear and G. Cullen, Eds.) 626-633, Electrochem. Soc., Pennington, NJ 08534 (1990)

52. Ahlgren, W., Jensen. J. and Olsen, G., Laser Assisted MOCVD of CdTe, *Proc. 11th. Conf. on CVD,* (K. Spear and G. Cullen, Eds.) 616-615, Electrochem. Soc., Pennington, NJ 08534 (1990)

53. Goossens, A., Kelder, E. and Schoonman, J., The Stabilization of Si Photoelectrodes by a Boron Phosphide (BP) Protective Optical Window, *Proc. 11th. Conf. on CVD,* (K. Spear and G. Cullen, Eds.), 567-572, Electrochem. Soc., Pennington, NJ 08534 (1990)

54. Chu, S., Chu, T., Han, K., Liu, Y. and Mantravadi, M.,

Chemical Vapor Deposition of Cadmium Telluride, *Proc. 10th Int. Conf. on CVD*, (G. Cullen, Ed.) 982-989, Electrochem. Soc., Pennington, NJ 08534 (1987)

7

THE CVD OF CERAMIC MATERIALS

1.0 INTRODUCTION

The term ceramics originally referred to oxides only. This is no longer the case as the meaning has been considerably broadened and now includes such inorganic compounds as borides, carbides, nitrides, oxides, silicides and compounds of Group VIb, i.e. the chalcogenides (sulfides, selenides and tellurides). In this chapter, the CVD of these ceramic materials will be reviewed. The number of ceramic compounds is very large and only those which have been the object of significant CVD investigation are reviewed here.

2.0 THE CVD OF BORIDES

2.1 General Characteristics and Properties

Boron forms stable borides with the transition metals. The most refractory of these compounds and those with the greatest potential interest are the borides of the elements of Groups IVa, Va and, to a lesser degree, VIa. They are shown in Table 1.

TABLE 1

The Refractory Metals in the Periodic Table
(atomic number in bracket)

	Group		
	IVa	Va	VIa
Period	4Ti (22)	V (23)	Cr (24)
Period	5Zr (40)	Nb (41)	Mo (42)
Period	6Hf (72)	Ta (73)	W (74)

These borides have a structure which is dominated by the boron configuration. This clearly favors the metallic properties such as high electrical and thermal conductivities and high hardness. Chemical stability, which is related to the electronic structure of the metal, decreases from the borides of the metals of Group IVa to those of Group VIa. Thus, the most stable borides are TiB_2, ZrB_2 and HfB_2. Properties of interest are summarized in Table 2 (1).

The borides listed in Table 2 can all be produced by CVD. With a few exceptions, they have found only limited industrial applications so far, in spite of their outstanding properties of hardness, erosion resistance and high temperature stability.

2.2 Boriding

Boriding by CVD is a relatively simple process wherein a layer of boron is deposited on a metal substrate, followed by heat treatment (2). The boron can be deposited by the hydrogen reduction of the chloride or by the decomposition of diborane (see Chapter Six, Section 2). During heat treatment, the metal boride is formed by interstitial diffusion of the boron, although the metal atoms will usually diffuse at a much lower rate than the boron since their atomic radii are considerably greater, i.e., 1.17 Å for boron, 2.0 Å for titanium and 2.16 Å for zirconium.

TABLE 2

Properties of the Borides

Boride	Density g/cm³	Melting Point°C	Hardness Kg/mm² (VHN50)	Electrical Resistiv µohm-cm	Thermal Conduc. w/cm.°C	Thermal Expans. ppm/°C (300-1000°C)
HfB_2	11.20	3250	2900	10-12	-	6.3-6.8
Mo_2B_5	7.48	2100	2350	18-40	-	-
NbB_2	7.21	3050	2200	12-65	0.17	5.5-9.2
TaB_2	12.60	3200	2500	14-68	0.11	5.8-7.1
TiB_2	4.52	2980	3370	9-15	0.25	6.6-8.6
W_2B_5	13.10	2300	2660	21-56	-	-
ZrB_2	6.09	3040	2300	7-10	0.25	6.6-6.8

The rate of boride formation by diffusion increases when going from the metals of Group IVa to those of Group VIa (3,4). The rate is also related to the stability of the boride; the most stable are those of Group IVa (TiB_2, ZrB_2 and HfB_2) which are the slowest to form. On the other hand, boriding of tungsten is a rapid process since the rate of diffusion of boron in tungsten with conversion of tungsten to its borides is an important consideration in the industrial production of boron filaments by CVD of boron on tungsten filaments.

Boriding of titanium by CVD in a chloride based system is more difficult since the titanium substrate is highly susceptible to HCl attack and diffusion is very slow. Boriding is used extensively on steel with high hydrogen dilution of the BCl_3 to prevent substrate attack. An iron boride is formed (5).

2.3 Direct Boride Deposition

Unlike boriding, direct boride deposition does not require a reaction with the substrate to form the boride. Both boron and the metal are supplied as gaseous compounds.

Borides of Group IVa: TiB_2, ZrB_2 and HfB_2 are easily deposited by the hydrogen reduction of the halides. A typical reaction is as follows:

[1] $TiCl_4 + 2BCl_3 \longrightarrow TiB_2 + 10HCl$

ZrB_2 and HfB_2 are deposited by similar reactions. These reactions take place over a pressure range of a few Torr to 1 atm. and a temperature range of 800-1100°C, in a hydrogen atmosphere (6,7). The Group IVa borides can also be deposited with diborane as boron source in a pressure range of a few Torr to 1 atm. as follows:

[2] $TiCl_4 + B_2H_6 \longrightarrow TiB_2 + 4HCl + H_2$

The applicable temperature range is lower than that of Reaction 1, 600-1000°C, but free boron tends to be incorporated in the deposit below 650°C (8). ZrB_2 is deposited by a similar reaction (9).

Borides of Group Va: The borides of Group Va, Nb_2 and TaB_2, are more difficult to deposit than those of Group IVa, since the incorporation of free metal in the deposit is difficult to avoid. However, pure deposits can be obtained by the co-reduction of the bromides at high temperatures (1500°C) and low pressure or by the coreduction of the chlorides if the molar gas mixture is preheated to 700-800°C just before entering the reactor (10,11,12). The incorporation of free metal can also be eliminated by using diborane as a boron source. Deposits of metal-free TaB_2 were obtained in this manner at 500-970°C (13).

Borides of Group VIa: As with the borides of Group Va, the incorporation of free metal in the Group VIa borides is difficult to avoid. Both tungsten and molybdenum borides are obtained at very high temperature regime by the hydrogen reduction of the mixed bromides (14). Boriding appears a more effective method to form these borides in thin layers. (See Section 2.2, above.)

In addition to the thermal CVD reactions described above, a glow discharge plasma at 480-650°C has been used to deposit TiB_2 from the mixed chlorides (15).

2.4 Applications

• Experimental TiB_2 coatings for cemented carbide cutting tools and other wear- and erosion-resistant applications (pumps, valves, etc.) (16,17)

• ZrB$_2$ coatings for solar absorption (18)

• TiB$_2$ coatings for electrodes for aluminum production (Hall-cell cathodes). TiB$_2$ has very high resistance to molten aluminum. It is easily wetted by the molten metal and good electrical contact is assured.

3.0 THE CVD OF CARBIDES

Carbides produced by CVD include the refractory metal carbides and two very important non-metallic carbides: boron carbide and silicon carbide. Carbides are very hard and wear resistant. They have very high melting point, are thermally stable and are generally chemically resistant (although not oxidation resistant). Their composition can vary over a wide range, and as a result their properties can vary considerably. Some carbides, such as B$_4$C, SiC, TiC and WC, are major industrial materials with numerous applications.

The refractory metal carbides consist of those of the nine transition elements of Groups IVa, Va and VIa and the 4th., 5th. and 6th. Periods shown in Table 1. Their structures increase in complexity with increasing group number. Thus the carbides of Group IVa are characterized by a single cubic monocarbide. In those of Group Va, a M$_2$C phase exists with a narrow composition range at room temperature adjacent to the monocarbide. The carbides of Group VIa are far more complex and have several compositions.

A general problem in the CVD of refractory carbides relates to potential oxygen contamination during deposition. Carbides can dissolve considerable quantities of oxygen by substitution for carbon. This is particularly true when the lattice is deficient in carbon (19). To avoid the deleterious effect of oxygen substitution, it is essential to maintain a deposition system that is completly free of oxygen. Likewise, hydrogen can dissolve readily in the defect carbides and, since many CVD reactions are carried out in hydrogen, this may easily occur. It may be necessary to vacuum anneal the coating to remove the hydrogen. Finally, composition uniformity can be a serious problem and the careful control of stoichiometry is necessary.

3.1 Boron Carbide (B$_4$C)

Boron carbide is a non-metallic compound with the theoretical stoichiometric formula B$_4$C. Stoichiometry however is rarely achieved and the compound is usually boron rich. It has a rhombohedral structure, a low density (2.51 g/cm^3), a high melting point (2450°C), a low thermal expansion (4.5 ppm/°C at 25°C) and is one of the hardest materials, next to diamond and cubic boron nitride with a Vickers hardness of 5000 kg/mm$^{2.}$ It is an electrical insulator (as opposed to the metal carbides) with a resistivity averaging 5x10^6 μohm-cm at 20°C. It has a thermal conductivity of 0.35 W/cm.°C at 25°C.

CVD Reactions: The following CVD reactions are commonly used to deposit boron carbide (20,21,22). All three reactions use excess hydrogen:

[1] $4BCl_3 + CH_4 + 4H_2 \longrightarrow B_4C + 12HCl$
(temperature range: 1200-1400°C, pressure: 10-20 Torr)

[2] $4BCl_3 + CH_3Cl + 5H_2 \longrightarrow B_4C + 13HCl$
(temperature range: 1150-1250°C, pressure: 10-20 Torr)

[3] $4BCl_3 + CCl_4 + 8H_2 \longrightarrow B_4C + 16HCl$
(temperature range: 1050-1650°C, pressure: to 1 atm.)

Boron carbide has also been deposited from diborane as a boron source in a plasma at 400°C as follows:

[4] $2B_2H_6 + CH_4 \longrightarrow B_4C + 8H_2$

Applications:

- Matrix material for ceramic composites (23)

- Free-standing shapes (24)

- Coating for nozzles, dressing sticks for grinding wheels

- Coating for neutron flux control in nuclear reactors

- Coating for shielding against neutron radiation

3.2 Chromium Carbide

Chromium carbide has three phases of which Cr_7C_3 is the most common. This phase is extremely resistant to corrosion and has the most oxidation resistant of all metal carbides (900°C).

It has a trigonal structure with a density of 6.9 g/cm³ and a melting point of 1782°C. Its thermal expansion is 10 ppm/°C at 25°C, its electrical resistivity 84 μohm-cm at 25°C and its thermal conductivity 0.11 W/cm.°C at 20°C.

CVD Reactions: A common deposition reaction combines the metal chloride with a hydrocarbon such as butane at an optimum deposition temperature of 1000°C (25). Other hydrocarbons can also be used.

Another useful reaction is the decomposition of the chromium dicumene $Cr[(C_6H_5)C_3H_7]_2$ in a temperature range of 300-550°C and at pressures of 0.5-50 Torr (26).

Applications:

- Coatings for combined corrosion- and wear-resistance

- Intermediate layer for tool steel coatings (27)

3.3 Hafnium Carbide

Hafnium carbide (HfC) is an extremely refractory compound with a melting point of 3890°C. It has a face-centered cubic structure (f.c.c. B1) and a density of 12.6 g/cm^3. Its composition rarely reaches stoichiometry and it is normally metal rich. It is very hard with a Vickers hardness of 2900 kg/mm^2. Its thermal conductivity is 0.08 W/cm.°C at 25°C ; its thermal expansion, 6.6 ppm/°C at 25°C and its electrical resistivity, 50 μohm-cm at 20°C. It is formed by CVD mostly on an experimental basis.

CVD Reactions: The most common deposition system is the reaction of the metal chloride with a hydrocarbon which can be propane (C_3H_8), propene (C_3H_6), toluene (C_7H_8) or methane (CH_4) as follows (28,29,30):

[1] $HfCl_4 + CH_4 \longrightarrow HfC + 4HCl$

The chloride is usually generated *in situ*. This reaction occurs over a wide range of temperature (900-1500°C) and pressure from 10 Torr to atmospheric. Whisker formation has been observed at 1 atm. and 1230°C (31).

Another deposition reaction is as follows:

[2] $HfCl_4 + CH_3Cl + H_2 \longrightarrow HfC + 5HCl$

(1200°C, 10-20 Torr)

Usually the more stoichiometric carbides are obtained at the lower pressure.

Applications:

• Oxidation resistant coatings for carbon-carbon composites (co-deposited with SiC) (32)

• Production of whiskers (with nickel catalyst) (30,31)

• Coating for superalloys (29)

3.4 Niobium Carbide

Niobium carbide (NbC), also known as columbium carbide, is a refractory material with very high melting point (3500°C). Its composition is usually metal rich ranging from $NbC_{0.76}$ to $NbC_{0.99}$. It has a face centered cubic (f.c.c. B1) structure with a density of 7.1 g/cm³. It is a hard compound (2400 kg/mm² VHN50) with a thermal expansion of 6.0 ppm/°C at 25°C, a thermal conductivity of 0.14 W/cm. °C at 25°C and an electrical resistivity of 60-150 μohm-cm at 20°C. It is being used as a CVD coating mostly on an experimental basis.

CVD Reactions: Niobium carbide may be obtained by carburization of the metal substrate according to the following reaction at low pressure and at temperatures above 2000°C (2):

[1] $CH_4 + Nb \longrightarrow NbC + 2H_2$

It can also be produced by the hydrogen reduction of the respective chlorides at 1500-1900°C, over a wide range of pressures (33):

[2] $NbCl_5 + CCl_4 + 4\ 1/2H_2 \longrightarrow NbC + 9HCl$

Applications:

- Hard coating for the protection of niobium metal

- As a carbonitride for superconductor applications (34,35)

3.5 Silicon Carbide

Silicon carbide (SiC) is a major industrial material with a considerable number of applications. CVD plays a major role in its development and production.

Of the several phases of SiC, the one of major interest here is beta-SiC which has a cubic zinc blend structure. The compound has low density (3.21 g/cm^3), a high melting point (ca. 2700°C, with dissociation starting at 2500°C), It has good chemical resistance, particularly to oxidation owing to the formation of a thin adherent and protective film of silicon dioxide on the surface. It has high resistance to radiation, an electrical resistivity of 10-100 ohm-cm at 25°C and a wide band gap (2.2 eV at 25°C) with useful semiconductor properties when suitably doped.

Finally, it has high strength and is very hard (2800 Kg/mm^2). It has relatively high thermal conductivity (1.25 W/cm.°C at 25°C) and low thermal expansion (3.9 ppm/°C at 20°C).

CVD Reactions: Most SiC deposition systems involve the Si-C-H-Cl chemical combination. A very commonly used reaction is the decomposition of methyl trichlorosilane (MTS) (36,37,38):

[1] $CH_3SiCl_3 \longrightarrow SiC + 3HCl$

This reaction is carried out in a temperature range of 900-1400°C (optimum 1100°C) and at total pressures of 10-50 Torr in a hydrogen atmosphere. The deposition rate and the crystallite size increase with increasing partial pressure of MTS. Other precursor combinations are: $SiCl_4/CH_4$, $SiCl_4/CCl_4$, SiH_2Cl_2/C_3H_8 and $SiHCl_3/C_3H_8$ (39).

Another common deposition system is based on the reaction of silane with a hydrocarbon such as propane or benzene in the following simplified reactions (36)(40):

[2] $3SiH_4 + C_3H_8 \longrightarrow 3SiC + 10H_2$

[3] $6SiH_4 + C_6H_6 \longrightarrow 6SiC + 15H_2$

The useful temperature range is lower than that of Reaction 1 with 800°C being typical. A pressure of approximately 10 Torr is typical, although atmospheric pressure can also be used (41). Plasma CVD has been used with Reactions 2 and 3 to deposit SiC at considerably lower temperatures (200-500°C) (42).

The decomposition of methyl silane (CH_3SiH_3) is used to produce an amorphous SiC at 800°C and a crystalline SiC at 900°C (43).

Applications:

- Coatings for susceptors and heating elements for epitaxial silicon deposition (44)

- Coatings for fusion reactor applications (45)

- Nuclear waste container coatings (43)

- High-power, high-frequency and high-temperature semi-conductor devices (39)

- Coatings for ceramic heat exchanger tubes

- Radiation sensors (amorphous SiC)

- Matrix in ceramic composites

- Oxidation resistant coatings for carbon-carbon composites

- Radiation-resistant semiconductor material

- Heteroepitaxial deposit on silicon

- Fibers and whiskers (46)

3.6 Tantalum Carbide

Tantalum carbide (TaC) is a refractory material which is structurally and chemically very similar to niobium carbide.

Its composition is usually metal rich ranging from $TaC_{0.58}$ to $TaC_{0.99}$. It has a face centered cubic structure (f.c.c. B1) with a density of 7.1 g/cm^3 and a very high melting point (3880°C). It has a

Vickers hardness of 2400 kg/mm² (VHN50), a thermal expansion of 5.5 ppm/°C at 25°C, a thermal conductivity of 0.22 W/cm.°C at 25°C and an electrical resistivity of 40-175 μohm-cm at 20°C. It is being used as a CVD coating mostly on an experimental basis.

CVD Reactions: TaC can be obtained by the carburization of a tantalum substrate or coating at low pressure and at temperatures above 2000°C (2):

[1] $CH_4 + Ta \longrightarrow TaC + 2H_2$

It can also be produced by the hydrogen reduction of the respective chlorides at 1150-1200°C and a pressure of 10-20 Torr:

[2] $TaCl_4 + CH_3Cl + H_2 \longrightarrow TaC + 5HCl$

The metal chloride is usually produced *in situ* at 550-600°C (see Chapter 4 Section 3.1)

Applications:

• Hard coating for the protection of tantalum metal

3.7 Titanium Carbide

Titanium (TiC) is a major industrial material produced extensively by CVD. It has a face-centered cubic structure (f.c.c. B1) with a very wide range of composition ($TiC_{0.47}$ to $TiC_{0.99}$). Cubic TiC is isomorphous with TiN and TiO. Thus oxygen and nitrogen as impurities, or as deliberate addition, can substitute for carbon to form binary and ternary solid solutions over a wide range of homogeneity. Titanium carbide is very refractory with a melting point of 3250°C and very hard (3200 Kg./mm²), with high strength and rigidity and outstanding wear resistance. It has a low coefficient of friction and resists cold welding. It oxidizes above 550°C. It has a thermal expansion of 7.6 ppm/°C in the 25-300°C range, a thermal conductivity of 0.17 W/cm·°C at 20°C and an electrical conductivity of 60-250 μohm-cm at 25°C.

CVD Reactions: The most common deposition system is the reaction of the metal chloride with a hydrocarbon as follows:

[1] $TiCl_4 + CH_4 \longrightarrow TiC + 4HCl$

This reaction is usually carried out at a temperature range of 850-1050°C in a hydrogen atmosphere and at pressure varying from less than 1 Torr to 1 atm. (47)(48). Above 1300°C, single crystal TiC is deposited (2). Other carbon sources such as toluene and propane have also been used (49). Reaction 1 is also used in a plasma at a lower temperature range (700-900°C) and lower pressure (1 Torr) (50).

TiC is also produced by reacting the chloride with a carbon substrate as follows:

[2] $TiCl_4 + C + 2H_2 \longrightarrow TiC + 4HCl$

This reaction occurs in excess hydrogen in the temperature range of 1750-1800°C.

Recent investigations with MOCVD have shown that TiC can be deposited at lower temperature (700°C) (51). Metallorganics that are used include the following:

- Tris-(2.2'-bipyridine) decomposed at 370-520°C

- Tetraneopentyl titanium decomposed at 150-300°C

- Dichlorotitacene, $(C_5H_5)_2TiCl_2$ (substrate temperature is 700°C)

Titanium carbonitride (TiC_xN_{1-x}) combines the wear properties of TiC with the low friction and oxidation and chemical resistance of TiN. It can be obtained by the following simplified reaction:

[3] $TiCl_4 + xCH_4 + \frac{1}{2}(1-x)N_2 + 2(1-x)H_2 \longrightarrow TiC_xN_{1-x} + 4HCl$

This reaction is carried out in a hydrogen atmosphere and at a temperature of approximately 1000°C.

If acetonitrile (CH_3CN) is used as a carbon and nitrogen source, the deposition temperature is greatly reduced and the process can be used to coat tool steel (52,53). The reaction is carried out at low pressure and at a temperature range of 700-900°C. It is as follows (in simplified form):

[4] $TiCl_4 + CH_3CN + 2\frac{1}{2}H_2 \longrightarrow TiCN + CH_4 + 4HCl$

Applications:

- Coatings for cutting and milling tools and inserts

- Coatings for stamping, chamfering and coining tools

- Ball-bearing coatings

- Coatings for extrusion and spray gun nozzles

- Coatings for pump shafts, packing sleeves and feed screws for the chemical industry

• Coatings for molding tools and kneading elements for plastic processing

3.8 Tungsten Carbide

Tungsten carbide is a complex system with two basic compositions which often occur together: W_2C and WC. The most common is the monocarbide which has two different structures: alpha-WC with a h.c.p. structure and beta-WC with an f.c.c (B1) structure. The melting point of WC is 2700°C and its density is 15.8 g/cm³. It has a Vickers hardness of 1800 kg/mm² which, as opposed to other carbides, is maintained at high temperature with little change. The hardness is still 1500 kg/mm² at 800°C. The thermal conductivity of WC is 0.28 W/cm.°C at 25°C; its thermal expansion is 4.5 ppm/°C at 20°C and its electrical resistivity 17 μohm-cm at 25°C. It is deposited by CVD mostly on an experimental basis.

CVD Reactions: Most reactions use the halides as a source of metal as follows (54, 55):

[1] $WCl_6 + CH_4 + H_2 \longrightarrow WC + 6HCl$

The temperature range for Reaction 1 is 670-720°C. Deposition is usually done at a pressure of a few Torr and in excess hydrogen.

Another reaction uses methanol as a carbon source over a wide range of pressure from a few Torr to 1 atm.:

[2] $WF_6 + CH_3OH + 2H_2 \longrightarrow WC + 6HF + H_2O$

Tungsten carbonyl is also used as a metal source at 350-400°C but carbon tends to remain incorporated in the structure (2). The reaction is as follows:

[3] $W(CO)_6 \longrightarrow W + 6CO$

Applications:

• Production of powder for hot pressing or hot isostatic pressing of high-precision tooling (see Chapter Twelve, Section 3.1)

• Coating of fine-porosity carbon for catalytic applications (56)

3.9 Zirconium Carbide

Zirconium carbide (ZrC) is a very refractory compound with a melting point of 3450°C. It has a face centered cubic (f.c.c. B1) structure with a density of 6.57 g/cm³. It has a Vickers hardness of 2000 kg/mm², a thermal conductivity of 0.20 W/cm.°C at 20°C, a thermal expansion of 6.0 ppm. at 20°C and an electrical resistivity of 57-75 μohm-cm. at 25°C. It is produced by CVD mostly on an experimental basis but also for nuclear applications. Like TiC, cubic ZrC has a variable composition and forms solid solutions with oxygen and nitrogen over wide ranges of stoichiometry.

CVD Reactions: A common reaction uses the bromide as a metal source at a temperature range of 1350-1550°C in an atmosphere of hydrogen and argon (57):

$$ZrBr_4 + CH_4 \longrightarrow ZrC + 4HBr$$

Zirconium carbide has also been deposited from the tetrachloride with methane or cyclopropane as the carbon source (58).

Applications:

● Coating for atomic fuel particles (thoria and urania) (59)

3.10 Miscellaneous Carbides

Many carbides, besides those listed above, can be and have been produced by CVD. These materials are generally of lesser interest at the present time and are listed here for reference purposes:

● Beryllium carbide, Be_2C, produced by the pyrolysis of metallo-organics (2)

● Germanium carbide, GeC, produced by the reaction of acetylene and germane (60)

● Molybdenum carbide, Mo_2C, produced by the decomposition of the carbonyl (61)

● Thorium carbide, ThC_2, deposited from the iodide (2)

● Uranium carbide, UC_2, deposited from the iodide (2)

● Vanadium carbide, VC, deposited from the chloride and methane (2)

4.0 THE CVD OF NITRIDES

The nitrides commonly produced by CVD include several refractory metal nitrides and two non-metallic nitrides of great industrial importance, boron nitride and silicon nitride. These two materials have very interesting properties and their development is proceeding at a rapid pace. CVD is a major factor in this growth.

In most reactions, ammonia is normally used as a source of nitrogen, rather than nitrogen. The reason is a favorable dissociation energy. To dissociate molecular nitrogen N_2 requires 227 kCal/mole. To completly dissociate NH_3 requires the following steps:

$$NH_3 \longrightarrow NH_2\text{-}H \quad (110 \text{ kCal/mole})$$

$$NH_2 \longrightarrow NH\text{-}H \quad (90 \text{ kCal/mole})$$

$$NH \longrightarrow N\text{-}H \quad (79 \text{ kCal/mole})$$

Although the total is greater than the required energy for N_2 dissociation, the barrier at each step is lower and becomes favorable kinetically. As shown by this example, the dissociation energy of potential reactants can be a very important factor that must be considered when selecting a reaction.

4.1 Aluminum Nitride

The development of the CVD of aluminum nitride (AlN) is relatively recent and stems from the stability of the material and its unusually high thermal conductivity (up to 3.2 W/cm.°C at 25°C for single crystal with no impurities). Aluminum nitride is a refractory material which sublimes above 2400°C. It has a hexagonal structure and low density (3.28 g/cm^3). Its thermal expansion is low (4.3 ppm/°C in the 25-400°C range) which matches that of silicon. It has a Vickers hardness of 1225 kg/mm^2. It is optically transparent in the visible and near IR range. It has piezoelectric characteristics. It is an electrical insulator with a resistivity $> 10^{14}$ ohm-cm. Its dielectric constant is 8.9 at 1 MHz and its energy gap is 6 eV. It is deposited by CVD both experimentally and on a production basis.

CVD Reactions: AlN is deposited by the following high temperature reactions in a hydrogen atmosphere at low pressure (1 Torr) with ammonia and either the chloride or the bromide as metal sources (62,63):

[1] $AlCl_3 + NH_3 \longrightarrow AlN + 3HCl$

(1000-1100°C)

[2] $AlBr_3 + NH_3 \longrightarrow AlN + 3HBr$

(900°C)

Reaction 2 is also used with a plasma at a deposition temperature of 200-800°C (42).

AlN is now often produced by MOCVD by reacting ammonia with trimethyl aluminum at low pressure (< 1 Torr) in the temperature range 900-1400°C (64,65):

[3] $(CH_3)_3Al + NH_3 \longrightarrow AlN + 3CH_4$

The pyrolysis of aluminum-nitrogen organic complexes such as diethyl aluminum azide $[(C_2H_5)_2AlN_3]$ is also used successfully at low deposition temperatures (450-870°C) (66).

Applications:

• Heat sink substrate and packaging material for electronic devices

• Passivation and dielectric layers

• High-frequency acoustic wave devices (piezoelectric)

4.2 Boron Nitride

Boron nitride has a hexagonal structure very similar to that of graphite which it resembles in many ways. It has a very large anisotropy in the crystal with resulting anisotropic properties. The material obtained by CVD is usually a turbostratic boron nitride with warped basal planes and lattice defects. CVD boron nitride is also known as pyrolytic boron nitride or PBN (67).

Like graphite, it is soft and lubricious and it has a low density (2.25 g/cm³). It is a very refractory material which sublimes at 3000°C and is very chemically resistant. It has a low thermal expansion (1.2 ppm/°C at 25°C in the ab direction) and a thermal conductivity of 0.62 W/cm.°C in the ab direction and 0.014 W/cm.°C in the c direction at 20°C. It is an excellent dielectric with an electrical resistivity of 10^{20} ohm-cm at 25°C.

Boron nitride with a cubic structure is obtained by very high pressure processing. This structure is similar to that of diamond (cubic or c-BN) with extreme hardness and chemical resistance. Efforts to synthesize it by CVD at low pressure are promising (see Reaction 2 below) (67).

CVD Reactions. The reaction of boron trichloride and ammonia is as follows (2, 68):

[1] $BCl_3 + NH_3 \longrightarrow BN + 3HCl$

At a deposition temperature of 1300°C., a low-density boron nitride is obtained (1.5 g/cm³). Density increases with increasing temperature and reaches 2.0 g/cm³ at 1600°C. Vapor phase precipitation can be a problem in the high-temperature range (69). A more convenient reaction uses boron fluoride (68):

[2] $BF_3 + NH_3 \longrightarrow BN + 3HF$

(1100-1200°C, 1 atm.)

Reaction 2 is used in an electron cyclotron (ECR) plasma to produce cubic-BN at 675°C on an experimental basis (67).

Low-temperature deposition is possible from diborane as a boron source (70):

[3] $B_2H_6 + 2NH_3 \longrightarrow 2BN + 6H_2$

(300-400°C, < 1 Torr)

Another useful deposition reaction is the decomposition of borazine. This is a condensation reaction which produces an amorphous BN with residual hydrogen incorporation (71):

[4] $B_3H_3N_3 \longrightarrow 3BN + 1½ H_2$

(700°C, < 1 Torr)

MOCVD has also been used with triethyl boron as the boron source in a hydrogen and argon atmosphere (72):

[5] $B(C_2H_5)_3 + NH_3 \longrightarrow BN + \text{hydrocarbons}$

(750-1200°C)

Applications:

- Passivation layer for microelectronic devices

- Sodium barrier in IC's

- Masks for X-ray lithography

- High temperature crucibles and boats

- Electromagnetic windows

4.3 Hafnium Nitride

Hafnium nitride (HfN) has a cubic structure (f.c.c. B1), a density of 13.8 g/cm^3 and is very refractory (melting point 3310°C) and very resistant to chemical attack. It is hard (up to 2100 kg/mm^2, VHN50). It has a thermal expansion of 6.9 ppm/°C (0- 1000°C), a thermal conductivity of 0.022 W/cm.°C and an electrical resistivity of 10 μohm-cm at 20°C. It is produced by CVD mostly on an experimental basis.

CVD Reactions: The following reaction is used in excess hydrogen, at a temperature range of 900-1300°C and at low pressure (ca. 10 Torr) (73):

$$2HfCl_4 + N_2 + 4H_2 \longrightarrow 2HfN + 8HCl$$

A similar reaction with ammonia as nitrogen source at 1100°C has a much higher deposition rate, owing to the high reactivity of the monatomic nitrogen released in the ammonia decomposition.

Applications:

- Coatings for cutting tools

- Tribological and corrosion resistant coatings

• Diffusion barriers for microelectronic devices

4.4 Niobium Nitride

Niobium nitride (NbN) has a cubic structure (f.c.c. B1), a melting point of 2330°C, a density of 8.3 g/cm³ and is very resistant to chemical attack. Its Vickers hardness is 1400 kg/mm². It has a thermal expansion of 10 ppm/°C (0-1000°C), a thermal conductivity of 0.04 W/cm.°C and an electrical resistivity of 200 μohm-cm at 20°C. It is a superconductor. It is produced by CVD mostly on an experimental basis.

CVD Reactions: The following reaction is used in excess hydrogen, at a temperature range of 1000-1100°C and low pressure (ca. 10 Torr):

$$NbCl_4 + NH_3 + 1/2H_2 \longrightarrow NbN + 4HCl$$

The reaction of the chloride with nitrogen has also been used (74,75). Lower deposition temperature is possible by MOCVD via the pyrolysis of a dialkylamide at 500-800°C (76).

Applications:

• As a potential superconductor coating (74)

4.5 Silicon Nitride

Silicon nitride (Si_3N_4) is a major industrial material which is produced extensively by CVD for electronic and structural applications. References to silicon nitride in the literature are very extensive.

Silicon nitride can be obtained as an amorphous material or in two hexagonal crystalline forms, alpha and beta, the latter being the high-temperature form. An irreversible transformation from alpha to beta occurs at 1600°C. Silicon nitride has a density of 3.18 g/cm³, a melting point of 1900°C, a low thermal expansion (2.5 ppm/°C over the range of 0-100°C) and a thermal conductivity of 0.45 W/cm.°C at 25°C. It is an insulator with an electrical resistivity of 10^{14} ohm-cm at 20°C, with very high dielectric strength. It has excellent mechanical properties.

CVD Reactions: Two reactions of major interest are employed to deposit silicon nitride. The first uses silicon tetrachloride ($SiCl_4$) with ammonia:

[1] $3SiCl_4 + 4NH_3 \longrightarrow Si_3N_4 + 12 HCl$

The optimum deposition temperature is 850°C. Pressure may be up to 1 atm.. A hydrogen or nitrogen atmosphere is used with very high ratio of N_2 to reactants (ca. 1000/1) (77,78,79).

The second reaction of major interest uses dichlorosilane (SiH_2Cl_2), also with ammonia:

[2] $3SiH_2Cl_2 + 4NH_3 \longrightarrow Si_3N_4 + 6HCl + 6H_2$

The range of deposition temperature is 755-810°C with a high dilution of nitrogen (80). When a high-frequency plasma (13.56 MHz) is used, the deposition temperature is lower (400-600°C) (81).

Another common deposition reaction, used widely in semiconductor processing, combines ammonia with silane as the silicon source:

[3] $3SiH_4 + 4NH_3 \longrightarrow Si_3N_4 + 12H_2$

Deposition temperature range from 700 to 1150°C and pressure up to 1 atm. Excess ammonia is used since it decomposes slower than silane. The ammonia-to-silane ratio should be greater than 10/1 over stoichiometric (78).

Reactions 1, 2 and 3, which all use ammonia, have a tendency to deposit silicon nitride with a high ratio of included hydrogen, especially at the lower temperatures and if a plasma is used. This tendency is often detrimental but it can be remedied, at least to some degree, by using nitrogen instead of ammonia:

[4] $3SiH_4 + 2N_2 \longrightarrow Si_3N_4 + 6H_2$

However, nitrogen has a far greater bonding energy than ammonia and is more difficult to dissociate into free atomic nitrogen active species. Consequently, the deposition rate is extremely slow. This can be offset by plasma activation with high frequency (13.56 MHz) or electron cyclotron resonance (ECR) plasmas (82,83,84). Plasma deposition lowers the content of included hydrogen.

Applications:

• Powders for hot pressing or sintering of high- strength refractory structural parts (pump bearings, heat exchangers, cutting tips, etc.)

• Crucibles for silicon single-crystal processing (85)

• Passivation layers, multilayer resist stacks, diffusion barriers, interlevel dielectrics, side wall spacers, trench masks, oxidation masks, etc., in semiconductor processing

• Whiskers for high strength reinforcements

• High temperature turbine blades

• Components for internal combustion engines

4.6 Titanium Nitride

Titanium nitride (TiN) is used extensively as a CVD coating mainly for wear- and erosion-resistant applications. TiN has a cubic structure (f.c.c. B1), a melting point of 2950°C, a density of 5.4 g/cm^3, a Vickers hardness of 2100 kg/mm^2. It is a chemically very stable material with a low coefficient of friction. It constitutes an excellent diffusion barrier especially for silicon. It has a thermal expansion of 9.5 ppm/°C in the 0-1100°C range, a thermal conductivity of 0.33 W/cm.°C at 25°C and an electrical resistivity of 135 μohm-cm at 20°C.

CVD Reactions: Two major reactions are used to deposit TiN, both based on titanium chloride as the metal source, as follows:

[1] $TiCl_4 + \frac{1}{2} N_2 + 2H_2 \longrightarrow TiN + 4HCl$

[2] $TiCl_4 + NH_3 + \frac{1}{2} H_2 \longrightarrow TiN + 4HCl$

The range of temperature for Reaction 1 is 900-1200°C with best results obtained at 1000°C. An argon diluent is used at pressures up to 1 atm. (86). Reaction 2 takes place at lower temperature (575-700°C) and is usually carried out at low pressure (0.5 Torr) with excess hydrogen (87).

Reaction 1 is also obtained in a high frequency plasma (13.56 MHz) at 1.1 Torr pressure and at a low deposition temperature of 500°C (88, 89, 90).

Applications:

• Wear- and erosion-resistant coatings on cemented carbides, either singly or in combination with TiC, TiCN and Al_2O_3.

• Coatings on tool steel for twist drills (89, 90)

• Diffusion barriers in semiconductor devices, between Si and Al, Ti and Pt, and Ag and Si (88)

4.7 Miscellaneous Nitrides

The following nitrides have been deposited mostly on an experimental basis (2, 78):

• Gallium nitride (GaN) from the ammonolysis of $GaCl_3$

• Germanium nitride (Ge_3N_4) made from the ammonolysis of $GeCl_4$

• Tantalum nitride (TaN) from the metal chloride reaction with nitrogen at 800-1500°C

• Zirconium Nitride (ZrN) from the metal chloride reaction with nitrogen at 1150-1200°C

5.0 THE CVD OF OXIDES

Oxides are the original and largest group of ceramic compounds, notable for their chemical inertness, good high temperature properties and resistance to oxidation. Most oxides have a high degree of ionic bonding since oxygen is the most electronegative divalent element. As a result, they generally have the characteristics of ionic crystals, i.e. optical transparency when pure, high electrical resistivity, low thermal conductivity, diamagnetism and chemical stability. There are notable exceptions; for instance some oxides are electrically conductive such as indium and tin oxides. Others, such as beryllium oxide, have high thermal conductivity .

There are many oxides since most metallic elements form stable compounds with oxygen and, of course, the number of mixed oxides is far greater. However, the CVD of many of these materials has yet to be investigated. Generally this area of CVD has lagged somewhat behind the CVD of other ceramics particularly in the development of optical applications where evaporation and sputtering processes were, until recently, used almost exclusively. This situation is changing and CVD is now fast becoming a major process for oxide deposition.

5.1 Aluminum Oxide (Alumina)

Alumina (Al_2O_3) is a widely used ceramic produced extensively by CVD. It has a rhombohedral structure, relatively low density (3.9 g/cm³), high strength and rigidity, good chemical and oxidation resistance, a melting point of 2015°C, a thermal expansion of 8.3 ppm/°C at 25-800°C, a thermal conductivity of 0.34 W/cm.°C at 25°C (fully dense material). It is a good electrical insulator with a resistivity of 10^{16} ohm-cm at 25°C and a dielectric constant of 8.5. It has a Vickers hardness of 1910 kg/mm². Its refractive index is 1.75 (corundum).

CVD Reactions: A common reaction for the deposition of alumina is the hydrolysis of the aluminum trichloride:

[1] $2AlCl_3 + 3H_2 + 3CO_2 \longrightarrow Al_2O_3 + 3CO + 6HCl$

This reaction, based on the water-gas reaction, takes place in excess hydrogen, at an optimum temperature of 1050°C and at low pressure (ca. 1 Torr) (2,91,92). It is the preferred reaction for tool coatings and electronic applications. The formation of alumina is dependent on the rate of formation of H_2O which is the rate-limiting factor. The control of H_2O formation is critical as H_2O and $AlCl_3$ react rapidly. If the concentration of H_2O is too high, gas phase precipitation results which leads to a powdery deposit. At 850°C, the alumina is amorphous. It becomes crystalline above 1000°C with a fine, uniform grain structure.

MOCVD is also used to deposit alumina at much lower temperatures than Reaction 1. A common reaction is the pyrolysis of an aluminum alkoxide such as aluminum isopropoxide, $Al(OC_3H_7)_3$. The formation of Al_2O_3 proceeds readily since the energy of the O-C bond (200-300 kJ/mol) is much lower than that of the Al-O bond (ca. 500 kJ/mol). No additional oxygen source is required as the ratio of Al/O in the precursor is 1/3. Decomposition occurs above 300°C, but it is necessary to operate above 900°C to obtain alpha-alumina (93). The decomposition of aluminum acetyl acetonate is also used (78).

The decomposition of aluminum alkyls, such as $(CH_3)_3Al$ or $(C_2H_5)_3Al$, in an oxidizing atmosphere such as O_2 or N_2O, produces alumina deposits in a temperature range of 250-500°C (78).

Applications:

• Coatings for carbide tools (usually with TiC and TiN underlayers) (94,95)

• Sealant coatings for plasma-sprayed oxides (96)

• Thin films in the fabrication of transistors (FET) and other semiconductor applications (92)

5.2 Chromium Oxide (Chromia)

Chromium oxide (Cr_2O_3) is a refractory material with excellent corrosion and oxidation resistance. It has an hexagonal structure, a density of 5.21 g/cm³ and a melting point of 2265°C. Its thermal conductivity at 25°C is 0.3 W/cm.°C, its thermal expansion is 9.0 ppm/°C in the range of 25-1100°C and its electrical resistivity is $1.3x10^3$ ohm-cm at 350°C. It is an optical material with an index of refraction of 2.55. It is produced by CVD mostly on an experimental basis.

CVD Reactions: Chromium oxide can be deposited by the decomposition of chromium acetyl acetonate, $Cr(C_5H_7O_2)_3$, in the 520-560°C temperature range (97). It can also be deposited by the decomposition of the carbonyl in an oxidizing atmosphere (CO_2 or H_2O) at low pressure (< 5 Torr) (2).

Applications:

• Intermediate layer in corrosion- and erosion-resistant applications

5.3 Hafnium Oxide (Hafnia)

Hafnium oxide (HfO_2) is a very refractory and stable oxide with extremely low vapor pressure, which is similar in structure and properties to zirconium oxide. It goes through a polymorphic phase change from a monoclinic to a tetragonal structure at 1615°C. This change is accompanied by a volume reduction of approximately 3.4%, resulting in the development of large internal stresses. To avoid this disruption, hafnia can be stabilized to an open cubic structure (fluorite) by the addition of a small amount (ca. 10%) of another oxide of cubic symmetry such as yttria, Y_2O_3. Stabilized hafnia is particularly susceptible to oxygen diffusion, through oxygen vacancies in the lattice.

Hafnia has a density of 10.0 g/cm³, a thermal expansion of 9.4 ppm/°C in a temperature range of 0-2000°C, a low thermal conductivity (0.016 W/cm.°C at 25°C) and is an electrical insulator with a resistivity of 10^9 ohm-cm at 20°C and a high dielectric constant. It has a Vickers hardness of 1500 kg/mm². It is deposited by CVD mostly on an experimental basis.

CVD Reactions: The most common deposition reaction uses hydrolysis of a metal halide such as $HfCl_4$ in excess hydrogen (the water-gas reaction) (32):

$$HfCl_4 + 2CO_2 + 2H_2 \longrightarrow HfO_2 + 2CO + 4HCl$$

Optimum deposition temperature is 1000°C and pressure is low (< 20 Torr). The chloride is a solid at room temperature and is preferably generated *in situ*.

Hafnia can also be deposited by the pyrolysis of a metallo-organic such as hafnium acetylacetonate, $Hf(C_5H_7O_2)_3$, at 400- 750°C or hafnium trifluoro-acetylacetonate, $Hf(C_5H_4O_2F_3)_4$ at 500- 550°C with helium and oxygen as carrier gases (98).

Applications:

- Diffusion barrier in semiconductor devices

- Oxidation resistant coatings

- Wire coating for emitters

5.4 Iron Oxide

The oxide of trivalent iron, Fe_2O_3, has a cubic structure, a density of 5.2 g/cm³, a melting point of 1565°C and a refractive index of 3.0. It is an optical material. It is formed by CVD mostly on an experimental basis.

CVD Reactions: Iron oxide is produced by the reaction of a halide such as iron trichloride with water at a temperature range of 800-1000°C and at low pressure (2):

$$2FeCl_3 + 3H_2O \longrightarrow Fe_2O_3 + 6HCl$$

It can also be deposited by MOCVD by the decomposition of the acetylacetonate, $Fe(C_5H_7O_2)_3$, at 400-500°C, or of the iron trifluoro-acetylacetonate $Fe(F_3C_5H_4)_3$ at 300°C in oxygen (78). Another MOCVD reaction uses a microwave plasma to decompose iron cyclopentadienyl, $(C_5H_5)_2Fe$, in an oxygen atmosphere in a temperature range of 300-500°C at low pressure (1-20 Torr) (99).

Applications:

- Beam splitter and interference layer in optical devices

- Detector for ethyl alcohol

5.5 Silicon Dioxide (Silica)

Silicon dioxide (SiO_2) is a major industrial material with many applications particularly in the semiconductor industry in the form of coatings which are mostly produced by CVD.

SiO_2 has an hexagonal structure and a low density of 2.20

g/cm^3. It has a melting point of 1610°C (the molten SiO_2 has very high viscosity). Its thermal expansion is very low (0.5 ppm/°C in the 0-150°C range). It has low thermal conductivity (0.12 W/cm.°C in the 0-150°C range). It is moderately hard (600-1000 kg/mm^2) and is an excellent electrical insulator with a resistivity of $1x10^{21}$ μohm-cm at 20°C. It has a low refractive index (1.45 for fused silica).

CVD Reactions: Many reactions are being used to deposit silicon dioxide, both experimentally and in production, and many more are being investigated in the laboratory. The selection depends on the application, the temperature limitations, the equipment available and other factors.

The most common deposition reactions are based on the combination of silane with various oxidizers either as thermal CVD or plasma CVD as follows (78,100,101):

[1] $SiH_4 + O_2 \longrightarrow SiO_2 + 2H_2$
        (in limited oxygen supply)

[2] $SiH_4 + 2O_2 \longrightarrow SiO_2 + 2H_2O$

Both reactions may take place simultaneously at a deposition temperature of 450°C and at atmospheric pressure. Reaction 1 can also take place in a plasma at 15-300 mTorr and in a temperature range of 200-300°C with a silane-to-oxygen ratio of 10/1 in a flow of argon or helium. These two reactions are used to produce doped silica by adding a doping gas to the stream such as diborane (B_2H_6) or phosphine (PH_3).

Another common deposition reaction uses carbon dioxide as the oxygen source in a plasma at temperatures ranging from 200 to 600°C and pressure usually less than 1 Torr (42).

Nitrous oxide (N_2O) is another common oxidizer which is used with silane in a plasma at 200-350°C, with ratios of N_2O to silane of 15/1 to 30/1 (102). It is also used with dichlorosilane in a deposition temperature range of 850-950°C and at low pressure (< 1 Torr) (103):

[3] $SiCl_2H_2 + 2N_2O \longrightarrow SiO_2 + 2HCl + 2N_2$

Considerable work is being done with the MOCVD of silica both experimentally and on a production basis. The most common MOCVD reaction is the decomposition of tetraethyl orthosilicate (TEOS) shown in simplified form as follows (2, 100):

[4] $Si(OC_2H_5)_4 \longrightarrow SiO_2 + 4C_2H_4 + 2H_2O$

This reaction normally takes place at 700°C and at low pressure (< 1 Torr). Doping is accomplished by the addition of arsine, diborane or phosphine. The addition of ozone (O_3) to Reaction 4 at atmospheric pressure provides films with excellent properties (104).

Silica is also deposited by the decomposition of diacetoxy-ditertiarybutoxy silane (DADBS) at 450-550°C and low pressure (< 1 Torr) and by the decomposition of octamethyl-cyclotetrasiloxane in ozone at 400°C at low pressure (100,105,106).

Applications:

- Passivation layers, surface dielectric and doping barriers in semiconductor designs

- Intermetallic dielectrics

- Diffusion sources

• Etch barriers

• Oxidation protection of stainless steel in nuclear reactors (107)

• Preparation of optical fibers

• Passivation layers in energy-saving architectural glass (E-glass)

5.6 Tantalum Oxide (Tantala)

Tantalum oxide (Ta_2O_5) has useful optical and dielectric properties. It is a fairly stable oxide with an orthorhombic structure, a melting point of 1870°C and a density of 8.27 g/cm^3. It is moderately hard with a Vickers hardness of 1400 kg/mm^2. It has a refractive index of 2.1-2.2. It is an electrical insulator with a high dielectric constant (25-35). It is deposited by CVD mostly on an experimental basis.

CVD Reactions: The hydrolysis of the halide is a common deposition reaction usually carried out in excess hydrogen and in the temperature range of 600-900°C (78):

[1] $2TaCl_5 + 2½ O_2 + 5H_2 \longrightarrow Ta_2O_5 + 10HCl$

MOCVD reactions are used increasingly such as the pyrolysis of tantalum ethylate, $Ta(OC_2H_5)_5$, in oxygen and nitrogen at 340- 450°C and at a pressure of < 1 Torr. This is followed by an annealing cycle at 600-900°C (108). Tantala is also deposited by the pyrolysis of the tantalum dichlorodiethoxy acetylacetonate at 300-500°C (109).

Applications:

- Dielectric for storage capacitors (MIS capacitors)

- Gate insulators in MOS devices

- Optical coatings; anti-reflection coatings; and coatings for hot mirrors

5.7 Tin Oxide

Tin oxide, SnO_2, has unusual physical properties. It is a good electrical conductor with an electrical resistivity of 0.6 ohm-cm at room temperature. It is highly transparent to the visible and highly reflective to the infra-red spectrum, with an refractive index of 2.0-2.1. It is a volatile oxide which decomposes above 1500°C, which is below its melting point (1630°C). It has several crystalline forms, the most common being tetragonal. Its density is 7.0 g/cm^3. It is deposited by CVD on a large industrial scale mostly for optical applications.

CVD Reactions: Tin oxide is deposited with the chloride as a metal source at 600-800°C, usually at low pressure (2):

[1] $SnCl_4 + O_2 + 2H_2 \longrightarrow SnO_2 + 4HCl$

Other reactions, which are becoming more common, use metallo-organic compounds such as tetramethyl tin which is oxidized at atmospheric pressure or at low pressure (< 1 Torr) in the range of 350-600°C, typically 470°C (110,111,112). Fluorine doping to modify the optical properties is accomplished by the addition of CF_3Br in the gas stream.

Another MOCVD precursor is dimethyl tin chloride which is

reacted with oxygen at 540°C in the following (simplified) reaction (113):

[2] $(CH_3)_2SnCl_2 + O_2 \longrightarrow SnO_2 + 2CH_3Cl$

Tin oxide can be readily doped with antimony and alloyed with indium oxide to increase the electrical conductivity.

Applications:

• Energy-saving coatings for plate glass (E-Glass) and light bulbs

• Transparent electrodes in photovoltaic cells

• Transparent heating elements

• Antistatic coatings

• Coatings for solar cells

• Oxygen sensors for air/fuel control in combustion engines (niobium oxide, Nb_2O_5, is also used for this application)

5.8 Titanium Oxide (Titania)

Titanium dioxide (TiO_2) is the most common of several known titanium oxides. It has a tetragonal structure, a melting point of 1850°C, a density of 4.25 g/cm³, a thermal expansion of 9.0 ppm/°C in the 0-1000°C range and a thermal conductivity of 0.08 W/cm.°C at 25°c. It is an electrical insulator with a dielectric constant of 85 (perpendicular to axis). It has a refractive index of

2.6-2.9 (in the rutile form). It is deposited by CVD on an experimental and production basis.

CVD Reactions: The oxidation of the chloride is a common deposition reaction. It takes place at 400-1000°C often in a helium atmosphere as follows (78):

$$TiCl_4 + O_2 + 2H_2 \longrightarrow TiO_2 + 4HCl$$

TiO_2 is now mostly deposited by the pyrolysis of a titanium alkoxide. Alkoxides are usually the preferred precursors since they have higher vapor pressure than other metallo-organics of titanium. A common reaction uses titanium ethoxide, $Ti(OC_2H_5)_4$, in an oxygen and helium atmosphere at 450°C; another uses titanium tetraisopropoxide, $Ti(OC_3H_7)_4$, with oxygen at 300°C and at low pressure (< 1 Torr) (114,115).

Applications:

• Applications are mostly optical such as high-index films in multilayer interference filters, antireflection coatings, optical wave-guides and photoelectrochemical cells

• Dielectric layers in thin-film capacitors

5.9 Zinc Oxide

Zinc oxide (ZnO) has useful piezoelectric properties. It has an hexagonal structure (wurtzite type) with a density of 5.66 g/cm³. It is relatively unstable and decomposes above 1700°C which is below its melting point (1975°C).

CVD Reactions: Zinc oxide is deposited by MOCVD, with an alkyl precursor such as dimethyl zinc and tetrahydrofuran (THF) (116):

$$(CH_3)_2Zn + C_4H_8O + 5H_2 \longrightarrow ZnO + 6CH_4$$

The deposition temperature range is 300-500°C, the partial pressure of the alkyl is 0.5-2.5 Torr and that of THF is 20-80 Torr.

Applications:

- Piezoelectric devices, transducers

- Coatings for photoconductive devices

- Coating for non linear resistors (varistors)

- Overvoltage protectors

5.10 Zirconium Oxide (Zirconia) and Yttrium Oxide (Yttria)

Zirconia (ZrO_2) is a very refractory material with a melting point of 2700°C. It is very similar to hafnia. It goes through three solid polymorphic phase changes from monoclinic to tetragonal at 1170°C and to cubic at 2370°C. The cubic phase has a fluorite crystal structure (open cubic). The monoclinic-to-tetragonal change is accompanied by a large volume reduction of approximately 7.5% resulting in considerable stresses. To avoid this phase change, zirconia can be stabilized in the cubic phase by the addition of a small amount of a divalent or trivalent oxide of cubic symmetry such as MgO, CaO or Y_2O_3. The additive oxide cation enters the crystal lattice and increases the ionic character of the metal-oxygen bonds. The cubic phase is not thermodynamically stable below approximately 1400°C for MgO additions, 1140°C for CaO additions, and below 750°C for Y_2O_3 additions. However, the diffusion rates for the cations are so low at these subsolidus temperatures that the cubic phase can easily be quenched and retained as a metastable phase.

Stabilized zirconia, with its simple open cubic structure, is particularly susceptible to oxygen diffusion, through oxygen vacancies. The addition of 9 mol% Y_2O_3 produces 4.1% vacancies and an ionic conductivity of 300-500 ohm-cm at 1000°C . This high electrical conductivity is used in oxygen sensing and in heating-element applications. Zirconia is produced by CVD mostly on an experimental basis.

Stabilized (cubic) zirconia has low thermal conductivity (0.03 W/cm.°C at 20°C) and high thermal expansion (10.2 ppm/°C from 25-800°C). It has high strength and rigidity and a Vickers hardness of 1700 kg/mm². Its refractive index is 2.17-2.20.

CVD Reactions: Unstabilized zirconia is deposited by the reaction of the metal halide with CO_2 and hydrogen (the water-gas reaction) at 900-1200°C (2):

[1] $ZrCl_4 + 2CO_2 + 2H_2 \longrightarrow ZrO_2 + CO + 4HCl$

Attempts to deposit yttria-stabilized zirconia by combining Reaction 1 and a similar hydrolysis of YCl_3 as source of yttrium at 700-1000°C were inconclusive (117). Codeposition from the chlorides in oxygen at 1100°C has been claimed (118).

Yttria-stabilized zirconia is also deposited by MOCVD (119). Deposition can be accomplished by the co-decomposition of the tetramethyl heptadiones of zirconium and yttrium, $Zr(C_{11}H_{19}O_2)_3$ and $Y(C_{11}H_{19}O_2)_3$, at 735°C. Deposition is also achieved by the decomposition of the trifluoro-acetylacetonates in a helium atmosphere above 300°C (120).

Applications:

- Electrolytes, oxygen sensors, fuel cells, electronic conduction coatings, furnace elements

- Piezoelectricity devices, PLZT ceramics

- High-temperature passivation of microelectronic devices

- Structural composites

5.11 Mixed Oxides and Glasses

Many mixed oxides can be produced by CVD and the potential number of combinations is of course very large. In this section, some of the most common mixed oxides are reviewed.

Titanates: The titanates are a large group of mixed oxides with unusual optical, electrical and mechanical properties. Many of these materials can be deposited by CVD, and CVD may soon become an economical production process. The most important titanates are as follows.

PZT (lead zirconate titanate) and PLZT (lead lanthanum zirconate titanate) combine ferroelectic, optical and electronic properties and are used in optoelectronic and piezoelectric devices. Powders for hot pressing produced by CVD are being investigated.

Lead titanate ($PbTiO_3$) is a ferroelectric material with unusual pyroelectric and piezoelectric properties. It is deposited by

MOCVD from ethyl titanate and lead vapor in oxygen and nitrogen at 500-800°C (121).

Another ferroelectric material is bismuth titanate ($Bi_4Ti_3O_{12}$) which is deposited from triphenyl bismuth, $Bi(C_6H_5)_3$ and titanium isopropoxide at low pressure (5 Torr) and at temperatures of 600-800°C (122).

Strontium titanate ($SrTiO_3$) has a large dielectric constant of 12, and a high refractive index with potential opto-electronic applications. It is deposited by MOCVD from titanium isopropoxide and a strontium beta-diketonate complex at 600-850°C and 5 Torr (123).

Magnesia Aluminate (Spinel): This compound ($MgAl_2O_4$) is deposited by combining the two metal chlorides with CO_2 and H_2 at 950°C as follows (124):

$$[1] \quad MgCl_2 + 2AlCl_3 + 4CO_2 + 4H_2 \longrightarrow MgO.Al_2O_3 + 4CO + 8HCl$$

A potential application is as an insulator coating on silicon in semiconductor devices.

Glasses: Glasses are oxides which have hardened and become rigid without crystallizing. The glassy structure consists of silica tetrahedra or other ionic groups that provide a solid, non-crystalline structure.

A widely used glass is phosphosilicate (PSG) which is used extensively in semiconductor devices as a passivation and planarization coating for silicon wafers. It is deposited by CVD by the reaction of tetraethylorthosilicate (TEOS): $(C_2H_5O)_4Si$, and trimethylphosphate: $PO(OCH_3)_3$, in a molecular ratio corresponding to a concentration of 5 to 7% P. Deposition temperature is usually 700°C and pressure is 1 atm.

Borophosphosilicate glass (BPSG) provides lower reflow temperature (800°C) than PSG and is used increasingly. It is deposited by reacting tetraethylorthosilicate, trimethylphosphite and trimethylborate with oxygen. Another reaction uses the hydrides i.e. silane (SiH_4), phosphine (PH_3) and diborane (B_2H_6) with nitrous oxide (N_2O). The respective amount of boron and phosphorus in the deposit is approximately 5% by weight.

These reactions take place by thermal CVD either at atmospheric or at low pressure. However, the use of plasma CVD at low pressure (3-10 Torr) is rapidly increasing due to the advantageous lower deposition temperature (355°C) (125,126). BPSG is used extensively as an interlayer dielectric on polysilicon.

5.12 Oxide Superconductors

The deposition of thin films of the high-temperature superconductor yttrium barium copper oxide, $YBa_2Cu_3O_7$, is obtained from the mixed halides, typically YCl_3, BaI_2 and $CuCl_2$, with O_2 and H_2O as oxygen sources. Deposition temperature is 870-910°C (127).

$YBa_2Cu_3O_7$ films are also obtained by MOCVD from a mixture of acetylacetonates (tetramethyl heptadionate) of yttrium, barium and copper, typically at a pressure of 5 Torr and at a deposition temperature of 825°C (128,129).

These precursor materials are readily prepared and are available commercially (130).

6.0 THE CVD OF SILICIDES

Silicides are useful compounds characterized by their refractoriness and high electrical conductivity. There are many silicides since silicon reacts with most metals and, in many cases, more than one silicide is formed. For instance, there are five known tantalum silicides.

The silicides of major industrial importance are the disilicides of the refractory metals: molybdenum, tantalum, titanium, tungsten, vanadium and zirconium (131,132). These compounds are of great interest particularly to the semiconductor industry because of their low electrical resistivity and their ability to withstand high processing temperatures. Silicide properties are listed in Table 3.

An older process to form silicides is "siliconizing" which is a relatively simple CVD process used to provide oxidation and chemical resistance to refractory metals. The siliconizing reaction uses the metal substrate such as Mo or Ti as the metal source. Silicon diffuses readily in these metals at relatively low temperatures. Thus the silicide may be formed either by a displacement or by a hydrogen-reduction reaction. A typical reaction is the siliconizing of molybdenum as follows:

[1] $2SiCl_4 + Mo + 2H_2 \longrightarrow MoSi_2 + 4HCl$

The deposition temperature is above 1200°C and the deposit usually consists of an outer layer of $MoSi_2$ and an intermediate layer of MoSi (2, 133). Such reactions are difficult to control and they often result in mechanical stresses and voids at the interface which may cause adhesion failure. For that reason, a preferred approach is the direct deposition of the silicide by reacting a gaseous silicon compound with a gaseous metal compound as shown in the following sections.

TABLE 3

Selected Properties of Silicides

Property	$MoSi_2$	$TaSi_2$	$TiSi_2$	WSi_2	$ZrSi_2$
Density g/cm³	6.24	9.08	4.10	9.75	4.90
Structure	tetra-gonal	hexa-gonal	ortho-rhombic	tetra-gonal	ortho-rhombic
Melting Point, °C	2050	2200	1540	2165	1600
Thermal Conductivity W/cm.°C	0.49	0.38	0.46	0.48	0.15
Thermal Expansion ppm/°C (0-1000°C)	8.4	9.0	10.7	7.0	8.7
Electrical Resistivity μohm-cm	40-100	35-70	13-16	30-100	35-40
Oxidation Resistance	excel-lent	poor	good	poor	poor

6.1 Molybdenum Disilicide

Molybdenum disilicide ($MoSi_2$) is unusual among silicides because of its outstanding oxidation resistance. In the presence of air at high temperature, $MoSi_2$ oxidizes to form an adherent, non-porous surface layer which has been tentatively identified as alpha-cristobalite as the principal constituent with other minor but essential phases of SiO_2 or Mo-Si-O ternary compounds. This surface layer acts as an efficient oxygen barrier and eliminates or, at least, considerably reduces further oxidation. Thus oxidation of $MoSi_2$ is minimized up to approximately 1900°C, which is close to its melting point.

$MoSi_2$ is usually deposited by the reaction of the metal halide with silane. The simplified reactions are as follows (134,135):

[1] $MoF_6 + 2SiH_4 \longrightarrow MoSi_2 + 6HF + H_2$

[2] $MoCl_5 + 2SiH_4 \longrightarrow MoSi_2 + 5HCl + 1\frac{1}{2} H_2$

Deposition temperature from the fluoride is 250-300°C at a pressure of < 2 Torr. Deposition temperature from the chloride is 650-950°C at low pressure. In Reactions 1 and 2 as well as in other silicide deposition reactions, the silane gas is usually highly diluted with an inert gas such as argon.

Also under investigation are MOCVD reactions from precursors having silicon-metal bonds such as $SiH_3Mo(CO)_3$ (136).

Applications:

- Conductive coatings in semiconductor devices

- Oxidation-resistant coatings

• Heating elements for high-temperature furnaces in oxidizing atmosphere

6.2 Tantalum Disilicide

Tantalum disilicide, $TaSi_2$, is very refractory and chemically resistant. It is deposited from the reaction of the chloride, $TaCl_5$, with silane (SiH_4) or dichlorosilane (SiH_2Cl_2), the latter precursor being preferred. The reaction takes place in a plasma as follows (137):

[1] $2SiH_4 + TaCl_5 \longrightarrow TaSi_2 + 5HCl + 1½ H_2$

Pressure is low (1-2 Torr) and amorphous deposits are obtained in the temperature range of 400-540°C.

Applications:

• Gate material in VLSI technology

6.3 Titanium Disilicide

Titanium disilicide ($TiSi_2$) has very low electrical resistivity and is a promising metallization material. It is deposited by the following reaction (138):

[1] $TiCl_4 + 2SiH_4 \longrightarrow TiSi_2 + 4HCl + 2H_2$

A similar reaction uses dichlorosilane as the silicon source:

[2] $TiCl_4 + 2SiH_2Cl_2 + 2H_2 \longrightarrow TiSi_2 + 8HCl$

Deposition temperature is 800°C and either atmospheric pressure or low pressure is used. This reaction can also be carried out in a plasma at very low pressure and at much lower temperature (450°C) (139).

A silicon substrate, such as the silicon wafer itself or a thin predeposited layer of silicon, may be used as the silicon source with the following reaction (140, 141):

[3] $TiCl_4 + Si \longrightarrow TiSi_2 + SiCl_4$ (3)

Deposition temperature is 850°C and pressure is < 1 Torr. The reaction is self-limiting since it relies on the diffusion of Si through the silicide.

Applications:

- Schottky barriers and ohmic contacts in IC's

- Replacement of doped silicon in MOS devices where silicon resistivity (300 μohm-cm) is too high

- General metallization

6.4 Tungsten Disilicide

Tungsten disilicide (WSi_2) is very refractory and stable with low resistivity. As with other silicides, a common deposition system uses silane as the silicon source with tungsten hexafluoride as follows (142):

[1] $2SiH_4 + WF_6 \longrightarrow WSi_2 + 6HF + H_2$

This reaction is usually carried out at 350°C and at low pressure (<1 Torr) with a high ratio of SiH_4 to WF_6

Reaction 1 is being replaced by the following reaction which uses dichlorosilane as the silicon source and provides better conformity and less cracking and peeling (143, 144):

[2] $2SiH_2Cl_2 + WF_6 + 3H_2 \longrightarrow WSi_2 + 2HCl + 6HF$

Deposition temperature is in the range of 400-650°C and pressure is < 1 Torr, with a high ratio of SiH_2Cl_2 to WF_6.

Another deposition reaction uses disilane as the silicon source at atmospheric pressure and at a deposition temperature of 290-300°C with nitrogen and hydrogen dilution as follows (145):

[3] $WF_6 + Si_2H_6 \longrightarrow WSi_2 + 6HF$

Applications:

   • Replacement for polysilicon gates and interconnects in MOS devices

   • Polycide structures (WSi_2 + polysilicon)

   • Adhesion layer with non-selective tungsten

6.5 Other Silicides

Several other silicides of potential interest are the object of active CVD development.

• Cobalt disilicide is deposited from the carbonyl $Co_2(CO)_8$ at 200°C in vacuum, followed by annealing at 700°C (146). It is also deposited from the pyrolysis of $SiH_3Co(CO)_4$ at 500°C (136).

• Iron disilicide is deposited from $(SiH_3)_2Fe(CO)_4$ at 500°C (136).

• Niobium silicide ($NbSi_3$) is deposited by the silane reaction with niobium chloride (147). It has an A15 structure and is a superconductor.

• Vanadium silicide, (VSi_2), is deposited from the chlorides (VCl_4 and $SiCl_4$) at 1000-1200°C and at 0.25 atm. (148).

7.0 THE CVD OF CHALCOGENIDES

The chalcogenides are binary compounds of a chalcogen (i.e. the elements of the Group VIb, oxygen, sulfur, selenium and tellurium), and a more electropositive element. This section covers the sulfides, selenides and tellurides. Oxides have been reviewed above in Section 5.

Most of the chalcogenides have useful optical characteristics and their applications are usually found in optics. CVD is a common production process.

In this section, three chalcogenides are reviewed: cadmium telluride (CdTe), zinc sulfide (ZnS) and zinc selenide (ZnSe), which are all produced on an industrial basis by CVD. Some aspects of the CVD of chalcogenides have also been reviewed in Chapter Six, Section 7.2.

7.1 Properties of the Chalcogenides

Selected properties of the chalcogenides are summarized in Table 4.

TABLE 4

Properties of Selected Chalcogenides

Property	CdTe	ZnSe	ZnS
Structure	cubic	cubic	hexagonal (alpha-ZnS)
Density, g/cm^3	5.85	5.27	4.09
Melting point, °C	1092	1520	1830
Hardness Knoop, kg/mm^2	45	100-150	250-350
Thermal conductivity W/cm.°C at 25°C	0.41	0.13	0.15
Thermal expansion ppm/°C at 0-450°C	6.2	7.6-8.2	7.5-7.8
Index of refraction at 10.6 μm	2.67	2.40	2.20
Useful bandpass absorption < 0.05 cm in μm	-	0.6-14	1-5,7-10
Mechanical properties	low	low	medium

7.2 Cadmium Telluride

Deposition of CdTe is obtained by the direct combination of the vapors of the two elements which are carried in a stream of hydrogen or helium. Cadmium vaporizes at 756°C and tellurium at 990°C (149). CdTe is also deposited by the following MOCVD reactions (151):

[1] $(C_2H_5)_2Te + Cd (g) + H_2 \longrightarrow CdTe + 2C_2H_6$

[2] $(C_2H_5)_2Te + (CH_3)_2Cd + nH_2 \longrightarrow CdTe +$ hydrocarbons

Reaction [1] takes place at 325-350°C. Reaction [2] is activated by an excimer laser (K_2F) at a substrate temperature of 150-250°C.

Applications:

• Thin film photovoltaic devices (CdTe is a direct bandgap semiconductor with a bandgap energy of 1.5 eV at room temperature)

7.3 Zinc Selenide

Zinc selenide (ZnSe) is deposited by the reaction of hydrogen selenide with zinc vapor transported in argon, at a deposition temperature of 700-750°C and at a pressure < 100 Torr (151):

[3] $Zn (g) + H_2Se (g) \longrightarrow ZnSe (s) + H_2 (g)$

Applications.

- Infra-red windows and other optical applications

7.4 Zinc Sulfide

Zinc sulfide (ZnS) has higher mechanical properties and better erosion resistance than zinc selenide but its optical transmission is not as good. It is deposited by a reaction similar to Reaction 3 with hydrogen sulfide as the hydrogen source (151,152)):

[4] $Zn \, (g) + H_2S \, (g) \longrightarrow ZnS \, (s) + H_2 \, (g)$

Deposition temperature is 600-800°C, pressure is 30-40 Torr and the ratio H_2S/Zn is 0.5-0.7.

Applications:

- Infra-red windows and other optical applications

- Windows for CO_2 lasers

REFERENCES

1. Pierson, H.O., A survey of the Chemical Vapor Deposition of Refractory Transition Metal Borides, in *Chemical Vapor Deposited Coatings*, 27-45, Am. Ceram. Soc. (1981)

2. Powell, C., Oxley, J. and Blocher, J.M., Jr., *Vapor Deposition,*

John Wiley & Sons, New York (1966)

3. Epik, A., *Poroshk. Metallurgiya*, 17(5):221-27 (1963)

4. Vandenbulcke, L. and Vuillard, G., *J. Electrochem. Soc.*, 123(2):278-285 (1976)

5. Singheiser, L., Wahl, G. and Hegewaldt, F., Deposition of Boron from $BCl_3/H_2/N_2$ Gas Mixtures on Steels with Low Alloy Content, *Proc. 9th. Int. Conf. on CVD*, (M. Robinson et al, Eds.) 625-638, Electrochem. Soc., Pennington, NJ 08534 (1984)

6. Caputo, A., Lackey, W. and Wright, I., Chemical Vapor Deposition of Erosion-Resistant TiB_2 Coatings, *Proc. 11th Int. Conf. on CVD*, (K. Spear and G. Cullen, Eds.) 782-799, Electrochem. Soc., Pennington, NJ 08534 (1990)

7. Pierson, H.O., Randich, E. and Mattox, D., The Chemical Vapor Deposition of TiB_2 on Graphite, *J. Less Common Metals*, 667(2):381-388, (Oct. 1979)

8. Pierson, H.O. and Mullendore, A., The Chemical Vapor Deposition of TiB_2 from Diborane, *Thin Solid Films*, 72(3):511-515, (15 Oct. 1980)

9. Randich, E., Low Temperature Chemical Vapor Deposition of TaB_2, *Thin Solid Films*, 72(3):517-522, (15 Oct. 1980)

10. Armas, C., Combescure, C and Trombe, F., *J. Electrochem. Soc.*, 123(2):308-310 (1973)

11. Motojima, S. and Kamiya, H., Chemical Vapor Deposition of the System Ti-Zr-B, *High Temp. High Press.*, 9:437-433 (Feb. 1982)

12. Motojima, S., Kito, K. and Sugiyama, K., Low Temperature

Deposition of TaB and TaB$_2$ by Chemical Vapor Deposition, *J. Nucl. Mat.*, 105(2-3):262-268 (Feb. 1982)

13. Randich, E., *Thin Solid Films*, 72:517-522, (1980)

14. Armas, B., *Rev. Int. Hte. Temp. Refract.*, 13(1):49-69 (1976)

15. Williams, L.M., Plasma Enhanced Chemical Vapor Deposition of Titanium Diboride Films, *Appl. Phys. Lett.* 2, 46-1:43-45 (Jan. 1985)

16. Caputo, A., Lackey, W., Wright I. and Angelini, P., Chemical Vapor Deposition of Erosion-Resistant TiB$_2$ Coatings, *J. Electrochem. Soc.*, 132(9):2274-2280, (Sept. 1985)

17. Graham, D., Wear Behavior of TiB$_2$ and Boron Coated Cemented Carbides, *Carbide Tool J.*, 18(5):35-43 (Sept. Oct. 1986)

18. Randich, E. and Allred, D., Chemical Vapor Deposited ZrB$_2$ as a Selective Solar Absorber, *Thin Solid Films,* 83:394-398 (1981)

19. Storms, E.K., *The Refractory Metal Carbides*, Academic Press, New York (1967)

20. Lartigue, S., Cazajous, D., Nadal, M. and Male, G., Study of Boron Carbide Vapor-Deposited under Low Pressure, *Proc. 5th. European Conf. on CVD* , (J. Carlsson and J. Lindstrom, Eds.) 403-410, Univ. of Uppsala, Sweden (1985)

21. Vandenbulcke, L. and Vuillard, G., Composition and Structural Changes of Boron Carbide Deposited by Chemical Vapor Deposition under Various Conditions of Temperature and Supersaturation, *J. Less-Common Met.*, 82:49-56 (Nov. Dec. 1981)

22. Mullendore, A., Chemical Vapor Deposition of Boron-based

Refractory Solids, *AIP Conf. Proc.*, 4-140, Am. Inst. of Physics, New York (1986)

23. Stinton, D., Ceramic Composites by Chemical Vapor Infiltration, *Proc. 10th Int. Conf. on CVD* (G. Cullen, Ed.) 1028-1040, Electrochem Soc., Pennington, NJ 08534 (1987)

24. Hayman, C., *Ceramic Surfaces and Surface Treatments*, (R. Morrell and M. Nicholas, Eds.), p.175, Brit. Ceram. Proc.34 (Aug. 1984)

25. Motojima, S. and Kuzuya, S., Deposition and Whisker Growth of Cr_7C_3 by CVD Process, *J. Crystal Growth* , 71(3):682-688 (1985)

26. Maury, F., Oquab, D., Morancho, R., Nowak, J. and Gauthier, J., Low Temperature Deposition of Chromium Carbide by LPCVD Process using Bis-Arene Chromium as Single Source, *Proc. 10th Int. Conf. on CVD*, (G. Cullen, Ed.) 1213-1219, Electrochem Soc., Pennington, NJ 08534 (1987)

27. Perry, A. and Horvath, E., Carbide Coating Interlayer on Steel, *Proc. 7th Int. Conf. on CVD* (T. Sedgwick and H. Lydtin, Eds.) 425-451, Electrochem Soc., Pennington, NJ 08534 (1979)

28. Hertz, D., Spitz, J. and Besson, J., Elaboration du Carbure de Hafnium par Depot Chimique en Phase Vapeur, *High Temp. High Press.*, 6:423-433 (1974)

29. Hakim, M., Chemical Vapour Deposition of Hafnium Nitride and Hafnium Carbide on Tungsten Wires, *Proc. 5th Int. Conf. on CVD* (J. Blocher et al, Eds), 634-649, Electrochem. Soc., Pennington, NJ 08534

30. Lackey, W., Hanigofsky, J. and Freeman, G.., Experimental Whisker Growth and Thermodynamic Study of the Hafnium-Carbon

System for Chemical Vapor Deposition Applications, *J. Amer. Ceram. Soc.*, 73(6):1593-98 (1990)

31. Futamoto, M., Yuito, I. and Kawabe, U., Hafnium Carbide and Nitride Whisker Growth by Chemical Vapor Deposition, *J. Cryst. Growth*, 61(1):69-74, (Jan. Feb. 1983)

32. Pierson, H.O., Sheek, J. and Tuffias, R., Overcoating of Carbon-Carbon Composites, *WRDC-TR-4045,* Wright-Patterson AFB, OH 45433 (Aug.1989)

33. Caputo, J., *Thin Solid Films*, 40-49 (1977)

34. Pike, G., Pierson, H.O., Mullendore, A. and Schirber, J., Superconducting Thin Film Niobium Carbonitrides on Carbon Fibers, *App. Polymer Symp. No. 29*, 71-81, John Wiley & Sons, New York (1976)

35. Brennfleck, K., Dietrich, M., Fitzer, E. and Kehr, D., Chemical Vapor Deposition of Superconducting Niobium Carbonitride Films on Carbon Fibers, *Proc. 7th Int. Conf. on CVD*, (T. Sedgwick and H. Lydtin, Eds.) 300- 314, Electrochem Soc., Pennington, NJ 08534 (1979)

36. Langlais, F. and Prebende, C., On the Chemical Process of CVD of SiC-based Ceramics from the Si-C-H-Cl System, *Proc. 11th. Int. Conf. on CVD*, (K. Spear and G. Cullen, Eds.) 686-695, Electrochem. Soc., Pennington, NJ 08534 (1990)

37. Schintlmeister, W., Wallgram, W. and Gigl, K., Deposition of CVD-SiC Coatings at Intermediate Coating Temperatures, *High Temp., High Press.*, 18(2): 211-222 (1986)

38. Fischman, G. and Petuskey, W., Thermodynamic Analysis and Kinetic Implications of Chemical Vapor Deposition of SiC from Si-

C-Cl-H Gas Systems, *J. Am. Ceram. Soc.*, 68(4):185- 190 (1985)

39. Furumara, Y., Doki, M., Mieno, F., Eshita, T., Suzuki, T. and Maeda, M., Heteroepitaxial beta-SiC on Si, *Proc. 10th. Int. Conf. on CVD* , (G. Cullen, Ed.) 435-444, Electrochem. Soc., Pennington, NJ 08534 (1987)

40. Komiyama, H., Oyamada, H., Tanaka, S. and Shimogaki, Y., Low Temperature Synthesis of SiC Films by Low Pressure Chemical Vapor Deposition, *Proc. 11th. Int. Conf. on CVD*, (K. Spear and G. Cullen, Eds.) 361-367, Electrochem. Soc., Pennington, NJ 08534 (1990)

41. Allendorf, M. and Kee, R., A Model of Silicon Carbide Chemical Vapor Deposition, *Proc. 11th. Int. Conf. on CVD* (1990), (K. Spear and G. Cullen, Eds.) 679-685, Electrochem. Soc., Pennington, NJ 08534 (1990)

42. Stinton, D., Besmann, T. and Lowden, R., Advanced Ceramics by Chemical Vapor Deposition Techniques, *Ceram. Bul.*, 67-2:350-355 (1988)

43. Angelini, P., Chemical Vapor Deposition of Silicon Carbide from Methylsilane and Coating of Nuclear Waste Ceramics, *Diss. Abstr. Int.*, 46(9):170, (Mar. 1986)

44. Brennfleck, K., Porsch, G., Reich, H. and Scheiffarth, H., CVD of SiC on an Industrial Scale, *Proc. 11th. Int. Conf. on CVD* , (K. Spear and G. Cullen, Eds.) 395-403, Electrochem. Soc., Pennington, NJ 08534 (1990)

45. Fukutomi, M., Kitajima, M., Okada, M. and Wanatabe, R., Silicon Carbide Coating on Molybdenum by Chemical Vapor Deposition and its Stability under Thermal Cycle Conditions, *J. Nucl. Mater.*, 87(1):107-116 (Nov. 1979)

46. Martineau, P., Pailler, R,. Lahaye, M. and Naslain, R., SiC Filament Matrix Composites Regarded as Model Composites, *J. Mater. Sci.*, 19(8):2749-2770 (Aug. 1984)

47. Aggour, L., Fitzer, E. and Schlichting, J., TiC Coating on Graphite, *Proc. 5th. Conf. on CVD*, (J. Blocher et al, Eds.) 600-610, Electrochem. Soc., Pennington, NJ 08534 (1975)

48. Dariel, M., Aparicio, R., Anderson, T. and Sacks, M., CVD of TiC_x on Refractory Materials, *Proc. 11th. Int. Conf. on CVD*, (K. Spear and G. Cullen, Eds.) 659-669, Electrochem. Soc., Pennington, NJ 08534 (1990)

49. Kim, D., Yoo, J. and Chun, J., Effect of Deposition Variables on the Chemical Vapor Deposition of TiC Using Propane, *J. Vac. Sci. Technol.* A4(2):219-221 (Mar. Apr. 1986)

50. Ikegawa, A., Tobioka, M., Doi, A. and Doi, Y., TiC and TiN Coated Cemented Carbides by RF Plasma Assisted CVD, *Proc. 5th. European Conf. on CVD* (J. Carlsson and J. Lindstrom, Eds.) 413-410, Univ. of Uppsala, Sweden (1985)

51. Huchet, G. and Teyssandier, F., Crystalline TiC Obtained at 700°C by MOCVD, *Proc. 11th. Int. Conf. on CVD,* (K. Spear and G. Cullen, Eds.) 703-809, Electrochem. Soc., Pennington, NJ 08534 (1990)

52. Pierson, H.O., Titanium Carbonitride Obtained by Chemical Vapor Deposition, *Thin Solid Films*, 40:41-47 (1977)

53. Chatterhee-Fischer, R. and Mayr, P., Investigations of TiCN-Layers Obtained at Moderate Temperatures, *Proc. 5th. European Conf. on CVD*, (J. Carlsson and J. Lindstrom, Eds.) 395-404, Univ. of Uppsala, Sweden (1985)

54. Teyssandier, F., Ducarroir, M. and Bernard, C., Investigation of the Deposition Conditions for Pure Tungsten Monocarbide, *Proc. 7th. Int. Conf. on CVD*, (T. Sedgwick and H. Lydtin, Eds.) 398-411, Electrochem Soc., Pennington, NJ 08534 (1977)

55. Roman, O.V., Kirilyuk, L. and Chernousova, S., Gas- Phase Precipitation of Tungsten Carbide Coatings, *Poroshk. Metall.*, (6):53-56 (1987)

56. Iizuka, M. and Komiyama, H., Preparation of Tungsten Carbides within the Porous Media as a Novel Catalyst by Chemical Vapor Deposition, *Proc. 11th. Int. Conf. on CVD* (K. Spear and G. Cullen, Eds.) 596-602, Electrochem. Soc., Pennington, NJ 08534 (1990)

57. Ogawa, T, Ikawa, K. and Iwamoto, K., Chemical Vapor Deposition of ZrC within a Spouted Bed by Bromide Process, *J. Nucl. Mater.*, 97(1-2):104-112 (Mar. 1981)

58. Hollabaugh, C., Wahman, L., Reiswig, R., White, R. and Wagner, P., Chemical Vapor Deposition of ZrC Made by Reactions of $ZrCl_4$ with CH_4 and with C_3H_6, *Nuc. Technol.*, 35(2):527-535 (Sept. 1977)

59. Ogawa, T., Ikawa, K., High Temperature Heating Experiments on Unirradiated ZrC-Coated Fuel Particles, *J. Nucl. Mater.*, 99(1):85-93 (July 1981)

60. Booth, D. and Voss, K., The Optical and Structural Properties of CVD Germanium Carbide, *J. Phys. (Orsay)*, 42, Suppl.4:1033-1036 (Oct. 1981)

61. Nutt, S. and Wawner, F., CVD Coating from Metal Carbonyls on SiC Filaments, *Proc. 10th. Int. Conf. on CVD* (G. Cullen, Ed.) 840-848, Electrochem. Soc., Pennington, NJ 08534 (1987)

62. Nickel, K., Riedel, R. and Petzow, G., Thermodynamic and Experimental Study of High-Purity Aluminum Nitride Formation from Aluminum Chloride by Chemical Vapor Deposition, *J. Amer. Ceram. Soc.*, 72(10):1804-1810 (1989)

63. Pauleau, Y., Bouteville, A., Hantzpergue, J., Remy, J. and Cachard, A., Composition, Kinetics and Mechanism of Growth of Chemical-Vapor-Deposited Aluminum Nitride Films, *J. Electrochem. Soc.*, 129(5):1045-1052 (May 1982)

64. Armas, B. and Combescure, C., Chemical Vapor Deposition of Si_3N_4 and AlN on Carbon Fibers, *Proc. 10th. Int. Conf. on CVD* (G. Cullen, Ed.) 1068-1069, Electrochem. Soc., Pennington, NJ 08534 (1987)

65. Susuki, M. and Tanji, H., CVD of Polycrystalline Aluminum Nitride, *Proc. 10th. Int. Conf. on CVD*, (G. Cullen, Ed.) 1089-1097, Electrochem. Soc., Pennington, NJ 08534 (1987)

66. Ho, K., Annapragada, A. and Jensen, K., MOCVD of AlN using Novel Precursors, *Proc. 11th. Int. Conf. on CVD* (K. Spear and G. Cullen, Eds.) 388-394, Electrochem. Soc., Pennington, NJ 08534 (1990)

67. Kempfer, L., The Many Faces of Boron Nitride, *Mater. Eng.*, 41-44 (Nov. 1990)

68. Pierson, H.O., Boron Nitride Composites by Chemical Vapor Deposition, *J. Composite Materials*, 9:228-240 (July 1975)

69. Gebhardt, J., CVD Boron Nitride Infiltration of Fibrous Structures, Properties of Low Temperature Deposits, *Proc. 4th. Int. Conf. on CVD* (G. Wakefield and J. Blocher, Eds.) , Electrochem. Soc., Pennington, NJ 08534

70. Adams, A. and Capio, C., *J. Electrochem. Soc.*, 127: 399 (1980)

71. Hirano, S., Yogo, T., Asada, S. and Naka, S., Synthesis of Amorphous Boron Nitride by Pressure Pyrolysis of Borazine, *J. Am. Ceram. Soc.* 72(1):66-70 (1989)

72. Nakamura, K., Preparation and Properties of Boron Nitride Films by Metal Organic Chemical Vapor Deposition, *J. Electochem. Soc.*, 133-6:120-1123 (1986)

73. Hakim, M., Chemical Vapor Deposition of Hafnium Nitride and Hafnium Carbide on Tungsten Wires, *Proc. 5th. Int. Conf. on CVD,* (J. M. Blocher Jr. et al, Eds.) 634-649, Electrochem. Soc. Pennington, NJ 08534 (1975)

74. Kieda, N., Mizutani, N. and Kato, M., CVD of 5a Group Transition Metal Nitrides, *Proc. 10th. Int. Conf. on CVD* (G. Cullen, Ed.) 1203-1209, Electrochem. Soc., Pennington, NJ 08534 (1987)

75. Funakubo, H., Kieda, N. and Mizutani, N., Preparation of Niobium Nitride Films by CVD, *Yogyo Kyokaishi*, Japan, 95(1):55-8 (1987)

76. Sugiyama, K., Pac, S., Tahashi, Y. and Motojima, S., *Proc. 5th. Int. Conf. on CVD*, (J. M. Blocher Jr. et al, Eds.), 147-154, Electrochem. Soc. Pennington, NJ 08534 (1975)

77. Bhat, D.G. and Roman, J. E., Morphological Study of CVD Alpha Silicon Nitride Deposited at one Atmosphere Pressure, *Proc. 10th. Int. Conf. on CVD* (G. Cullen, Ed.) 579-585, Electrochem. Soc., Pennington, NJ 08534 (1987)

78. Kern, W. and Ban, V.S., Chemical Vapor Deposition of Inorganic Thin Films, in *Thin Film Processes*, (J.. Vossen and W.

Kern, Eds.), Academic Press, New York (1978)

79. Kim, J., Yi, K. and Chun, J., The Effects of Deposition Variables in the Chemical Vapor Deposition of Si_3N_4, *Proc. 5th. European Conf. on CVD* (J. Carlsson and J. Lindstrom, Eds.) 482-491, Univ. of Uppsala, Sweden (1985)

80. Kaplan, W. and Zhang, S., Determination of Kinetic Parameters of LPCVD Processes from Batch Depositions, Stoichiometric Silicon Nitride Films, *Proc. 11th. Int. Conf. on CVD* (K. Spear and G. Cullen, Eds.), 381-387, Electrochem. Soc., Pennington, NJ 08534 (1990)

81. Marks, J., Witty, D., Short, A., Laford, W. and Nguyen, B., Properties of High Quality Nitride Films by Plasma Enhanced Chemical Vapor Deposition, *Proc. 11th. Int. Conf. on CVD* (K. Spear and G. Cullen, Eds.), 368-373, Electrochem. Soc., Pennington, NJ 08534 (1990)

82. Chang, M., Wong, J. and Wang, D., Low Stress, Low Hydrogen Nitride Deposition, *Solid State Technol.*, 193- 195, (May 1988)

83. Manabe, Y. and Yamazaki, O., Silicon Nitride Thin Films Prepared by ECR Plasma CVD, *Proc. 10th. Int. Conf. on CVD* (G. Cullen, Ed.), 885-893, Electrochem. Soc., Pennington, NJ 08534 (1987)

84. Tsu, D.V. and Lucovsky, G., Silicon Nitride and Silicon Diimide Grown by Remote Plasma Enhanced Chemical Vapor Deposition, *J. Vac. Sci. Technol.*, A4(3-1):480-485, (May-June 1986)

85. Shinko, J.S. and Lennartz, J.W., CVD Silicon Nitride Crucibles Produced Using SPC Techniques to Correlate and Optimize Processing Variables, *Proc. 10th. Int. Conf. on CVD* (G. Cullen, Ed.) 1106-1117, Electrochem. Soc., Pennington, NJ 08534 (1987)

86. Bhat, D.G., A Thermodynamic and Kinetic Study of CVD TiN Coating on Cemented Carbide, *Proc. 11th. Int. Conf. on CVD* (K. Spear and G. Cullen, Eds.) 648-655, Electrochem. Soc., Pennington, NJ 08534 (1990)

87. Sherman, A. and Raaijmakers, J., Step Coverage of Thick, Low Temperature LPCVD Titanium Nitride Films, *Proc. 11th. Int. Conf. on CVD*, (K. Spear and G. Cullen, Eds.) 374-380, Electrochem. Soc., Pennington, NJ 08534 (1990)

88. Ianno, N.J., Ahmed, A.U. and Englebert, D.E., Plasma-Enhanced Chemical Vapor Deposition of TiN from $TiCl_4/N_2/H_2$ Gas Mixtures, *J. Electrochem. Soc.*, 136-1 (Jan. 1989)

89. Shizhi, L., Cheng, Z., Yulong, S., Xiang, X., Wu, H., Yan, X. and Hongshun, Y., The Deposition of TiN Coatings by Plasma Chemical Vapor Deposition and its Applications, *Proc. 10th. Int. Conf. on CVD* (G. Cullen Ed.) 1233-1243, Electrochem. Soc., Pennington, NJ 08534 (1987)

90. Mayr, P. and Stock, H.R., Deposition of TiN and Ti(O,C,N) Hard Coatings by a Plasma-Assisted Chemical Vapor Deposition Process, *J. Vac. Sci. Technol.*, 4- (6):2726-2730 (Nov. Dec. 1986)

91. Alten, H., Stjernberg, K. and Lux, B., Influence of Oxygen and Water Traces Contained in the Reaction Gases Used for Al_2O_3 Deposition by CVD, *Proc. 5th. European Conf. on CVD* (J. Carlsson and J. Lindstrom, Eds.), 388- 393, Univ. of Uppsala, Sweden (1985)

92. Choi, S., Kim, C., Kim, J. and Chun, J., Nucleation and Growth of Al_2O_3 on Si in the CVD Process, *Proc. 9th. Int. Conf. on CVD*, (M. Robinson et al, Eds.), 233-241 Electrochem. Soc., Pennington, NJ 08534 (1984)

93. Pauer, G., Altena, H. and Lux, B., Al_2O_3 CVD with Organic Al-Donors, *Int. J. of Refract. & Hard Metals*, 5-(3):165-170 (Sept. 1986)

94. Colmet, R. and Naslain, R., Chemical Vapor Deposition of Alumina from $AlCl_3$, H_2 and CO_2 Mixtures, *Wear* 80(2): 221-231, (Aug. 16, 1982)

95. Hara, A., Alumina Coated Cutting Tools Yield Excellent Results in High-Speed Machining, *S. Afr. Mach. Tool Rev.*, 16(7)- 5.7.9. 11-12 (July 1983)

96. Mantyla, T., Vuoristo, P. and Kettunen, P., Chemical Vapor Deposition of Plasma Sprayed Oxide Coatings, *Thin Solid Films,* 118(4):437-444, (24 Aug. 1984)

97. Idrissi Chbihi, H., CVD of Chromium Oxide Coatings on Metal Substrates from Chromium Acetylacetone, *Proc. 4th. Int. Conf. on Surface Modification Technologies,* Societe Francaise de Metallurgie, Paris, France (1990)

98. Balog, M., Schieber, M., Michman, M. and Patai, S., Chemical Vapor Deposition and Characterization of HfO_2 Films from Organo-Hafnium Compounds, *Thin Solid Films*, 41:247-259 (1977)

99. Yu-Sheng, S. and Yuying, Z., Fe_2O_3 Ultrafine Particle Thin Films are Prepared by MO-CVD Method, *Proc. 10th. Int. Conf. on CVD* (G. Cullen, Ed.), 904-911, Electrochem. Soc., Pennington, NJ 08534 (1987)

100. *Handbook of Thin-Film Deposition Processes and Techniques,* (K. Schuegraf, Ed.) , Noyes Publications, Park Ridge, NJ (1988)

101. Blum, J., Chemical Vapor Deposition of Dielectrics, a Review, *Proc. 10th. Int. Conf. on CVD* (G. Cullen, Ed.) 476-491, Electrochem. Soc., Pennington, NJ 08534 (1987)

102. Kumagai, H., Plasma Enhanced CVD, *Proc. 9th. Int. Conf. on CVD* (M. Robinson et al, Eds.) 189-204, Electrochem. Soc., Pennington, NJ 08534 (1984)

103. Hosaka, T., Characteristics of Thin CVD-SiO_2 Film on Tungsten Polycide, *Proc. 10th. Int. Conf. on CVD* (G. Cullen, Ed.) 518-542, Electrochem. Soc., Pennington, NJ 08534 (1987)

104. Fujino, K., Nishimoto, Y., Tokumasu, N. and Maeda, K., Silicon Dioxide Deposition by Atmospheric Pressure and Low-Temperature CVD Using TEOS and Ozone, *J. Electrochem. Soc.*, 137-(9):2883-2888 (Sept. 1990)

105. Smolinsky, G., LPCVD of SiO_2 Films Using the New Source Material DADBS, *Proc. 10th. Int. Conf. on CVD* (G. Cullen, Ed.) 490-496, Electrochem. Soc., Pennington, NJ 08534 (1987)

106. Nishimoto, Y., Tokumasu, N., Fujino, K. and Maeda, K., New Low Temperature APCVD Method using Polysiloxane and Ozone, *Proc. 11th. Int. Conf. on CVD* (K. Spear and G. Cullen, Eds.) 410-417, Electrochem. Soc., Pennington, NJ 08534 (1990)

107. Bennett, M.J., Houlton, M.R., Moore, D.A., Foster, A.I. and Swidzinski, M., Development of a Chemical Vapor Deposition Silica Coating for UK Advance Gas-Cooled Nuclear Reactor Fuel Pins, *Nucl. Technol.*, 66(3):518- 522 (Sept. 1984)

108. Zasima, S., Furuta, T. and Yasuda, Y., Preparation and Properties of Ta_2O_5 Films by LPCVD for ULSI Applications, *J. Electrochem. Soc.*, 137-4:1297-1300 (1990)

109. Kaplan, E., Balog, M. and Frohman-Bentchowsky, D., *J. Electrochem. Soc.*, 124:4541 (1977)

110. Borman, C.G. and Gordon, R.G., Reactive Pathways in the

Chemical Vapor Deposition of Tin Oxide Films by Tetramethyltin Oxidation, *J. Electrochem. Soc.*, 136-12: 3820-3828 (1989)

111. Wan, C.F., McGrath, R.D., Keenan, W.F. and Franck, S.N., LPCVD of Tin Oxide from Tetramethyltin and Oxygen, *J. Electrochem. Soc.*, 136-5:1459-1463 (1989)

112. Gordon, R.C., Proscia, J., Ellis, F.R. and Delahoy, A.E., Textured Tin Oxide Films Produced by Atmospheric Pressure Chemical Vapor Deposition from Tetramethyltin and Their Usefulness in Producing Light Trapping in Thin-Film Amorphous Silicon Solar Cells, *Solar Energy Mater.*, 18:263-281 (1989)

113. Adachi, R. and Mizuhashi, M., CVD SnO_2 Films from Dimethyl Tin Dichloride, *Proc. 10th. Int. Conf. on CVD* (G. Cullen, Ed.) 999-1007, Electrochem. Soc., Pennington, NJ 08534 (1987)

114. Siefering, K.L. and Griffin, G.L., Growth Kinetics of CVD TiO_2: Influence of Carrier Gas, *J. Electrochem. Soc.*, 137- 4:1206-1208 (1990)

115. Balog, M., Schieber, M., Patai, S. and Michman, M., Thin Films of Metal Oxides on Silicon by Chemical Vapor Deposition with Organometallic Compounds, *J. of Crystal Growth*, 17:298-301 (1972)

116. Souletie, P. and Wessels, B.W., Growth Kinetics of ZnO Prepared by Organometallic Chemical Vapor Deposition, *J. Mater. Res.*, 3(4):740-744 (Jul/Aug 1988)

117. Lin, Y., Fransen, P., de Vries, K. and Burggraaf, A., Experimental Study on CVD Modification of Ceramic Membranes, *Proc. 11th. Int. Conf. on CVD* (K. Spear and G. Cullen, Eds.) 539-545, Electrochem. Soc., Pennington, NJ 08534 (1990)

118. Wahl, G., Schlosser, S. and Schmederer, F., *Proc. 8th. Int.*

Conf. on CVD, (J. M. Blocher, Jr. et al, Eds.) 536, Electrochem. Soc., Pennington, NJ 08534 (1981)

119. Sugiyama, K., Coating of Hard Refractory Materials by CVD, in *Fine Ceramics*, (S. Saito, Ed.) 62-71, Elsevier, New York (1988)

120. Balog, M., Schieber, M., Michman, M. and Patai, S., Zirconium and Hafnium Oxides by Thermal Decomposition of Zirconium and Hafnium Diketonate Complexes in the Presence and Absence of Oxygen, *J. Electrochem. Soc.*, 126(7):1203-1207 (1979)

121. Yoon, S. and Kim, H., Preparation and Deposition Mechanism of Ferroelectric $PbTiO_3$ Thin Films by Chemical Vapor Deposition, *J. Electrochem. Soc.*, 135-12:3137-3140 (1988)

122. Wills, L.A. and Wessels, B.W., Deposition of Ferroelectric $Bi_4Ti_3O_{12}$ Thin Films, *Proc. 11th. Int. Conf. on CVD* (K. Spear and G. Cullen, Eds.) 141-147, Electrochem. Soc., Pennington, NJ 08534 (1990)

123. Feil, W.A., Wessels, B.W., Tonge, L.M. and Marks, T.J., Chemically Vapor Deposited Strontium Titanate Thin Films and Their Properties, *Proc. 11th. Int. Conf. on CVD* (K. Spear and G. Cullen, Eds.) 148-1514, Electrochem. Soc., Pennington, NJ 08534 (1990)

124. Kavahara, K., Fukane, K., Inoue, Y., Taguchi, E. and Yoneda, K., CVD Spinel on Si, *Proc. 10th. Int. Conf. on CVD*, (G. Cullen, Ed.) 588-601, Electrochem. Soc., Pennington, NJ 08534 (1987)

125. Law, K., Wong, J., Leung, C., Olsen, J. and Wang, D., Plasma-Enhanced Deposition of Borophosphosilicate Glass Using TEOS and Silane Sources, *Solid State Technology*, 60-62, (April 1989)

126. Naimpally, A., A Study of the Annealing Processes for Doped

Glasses Deposited by the Chemical Vapor Deposition Process, *SAMPE J.*,24-5, (Sept/Oct 1988)

127. Ottosson, M., Harsta, A. and Carlsson, J., Thermodynamic Analysis of Chemical Vapor Deposition of $YBa_2Cu_3O_7$ from Different Halide Precursors, *Proc. 11th. Int. Conf. on CVD* (K. Spear and G. Cullen, Eds.) 180-187, Electrochem. Soc., Pennington, NJ 08534 (1990)

128. Thomas, O., Mossang, E., Pisch, A., Weiss, F., Madar, R. and Senateur, P., MOCVD of $YBa_2Cu_3O_7$ Thin Films, *Proc. 11th. Int. Conf. on CVD* (K. Spear and G. Cullen, Eds.) 219-228, Electrochem. Soc., Pennington, NJ 08534 (1990)

129. Lackey, W.J., Hanigofsky, J.A., Shapiro, M.J. and Carter, W.B., Preparation of Superconducting Wire by Deposition of $YBa_2Cu_3O_7$ onto Fibers, *Proc. 11th. Int. Conf. on CVD* (1990), (K. Spear and G. Cullen, Eds.) 195-210, Electrochem. Soc., Pennington, NJ 08534 (1990)

130. Barron, A.R., Group IIa Metal-Organics as MOCVD Precursors for High T_C Superconductors, *The Strem Chemiker*, XIII- 1, Strem Chemicals, Newburyport, MA 01950-4098 (Oct. 1990)

131. Murarka, S.P., *Silicides for VLSI Applications*, Academic Press (1983)

132. McLachlan, D.R., Refractory Metal Silicides, *Semiconductor International* (Oct. 1984)

133. Bernard, C., Madar, R. and Pauleau, Y., Chemical Vapor Deposition of Refractory Metal Silicides for VLSI Metallization, *Solid State Technology*, 79-84 (Feb. 1989)

134. West, G.A. and Beeson, K.W., Chemical Vapor Deposition of Molybdenum Silicide, *Proc. 10th. Int. Conf. on CVD* (G. Cullen, Ed.) 720-735, Electrochem. Soc., Pennington, NJ 08534 (1987)

135. Gaczi, P.J. and Reynolds, G.J., Identification of Reaction Products in the Low-Pressure Chemical Vapor Deposition of Molybdenum Silicide, *J. Electrochem. Soc.*, 136-9: 2661-2666 (Sept. 1989)

136. Aylett, B.J. and Tannahill, A.A., Chemical Vapour Deposition of Metal Silicides from Organometallic Compounds with Silicon-Metal Bonds, *SIRA Int. Seminar on Thin Film Preparation and Processing Technolgy*, Brighton UK (March 1985)

137. Hieber, K., Stolz, M. and Wieczorek, C., Plasma-Enhanced Chemical Vapor Deposition of $TaSi_2$, *Proc. 9th. Int. Conf. on CVD* (M. Robinson et al, Eds.) 205-2122, Electrochem. Soc., Pennington, NJ 08534 (1984)

138. Mastromatteo, E., Regolini, J.L., d'Anteroches, C., Dutartre, D., Bensahel, D., Mercier, J., Bernard, C.and Madar, R., Selective Chemical Vapor Deposition of $TiSi_2$, *Proc. 11th. Int. Conf. on CVD* (K. Spear and G. Cullen, Eds.), 459-466, Electrochem. Soc., Pennington, NJ 08534 (1990)

139. Hara, T., Ishizawa, Y. and Wu, H., Deposition and Properties of Plasma-Enhanced CVD Titanium Silicide, *Proc. 10th. Int. Conf. on CVD* (G. Cullen, Ed.), 867-876, Electrochem. Soc., Pennington, NJ 08534 (1987)

140. Schrey, F., Gallagher, P. and Levy, R., Selective Silicide or Boride Film Formation by Reaction of Vapor Phase $TiCl_4$ with Silicon or Boron, *J. Electrochem. Soc.*, 1137-5:1647-1649 (May 1990)

141. Engqvist, J. and Carlsson, J., Initial Stages of Growth During CVD of TiSi^2 on Unpatterned and Patterned Si Substrates, *Proc. 11th. Int. Conf. on CVD* (K. Spear and G. Cullen, Eds.), 448-452, Electrochem. Soc., Pennington, NJ 08534 (1990)

142. Sherman, A., *Chemical Vapor Deposition for MicroElectronics*, Noyes Publications, Park Ridge, NJ (1987)

143. Rode, E.J., Harshbarger, W.R. and Watson, L., Investigation of Gas Reactions during CVD of WSi$_2$, *Proc. 10th. Int. Conf. on CVD* (G. Cullen, Ed.), 711-719, Electrochem. Soc., Pennington, NJ 08534 (1987)

144. Hara, T., Miyamato, T., Hagiwara, H., Bromley, E. and Harshbarger, W., Composition of Tungsten Silicide Films Deposited by Dichlorosilane Reduction of Tungsten Hexafluoride, *J. Electrochem. Soc.*, 137-9: 2955-2959 (Sept. 1990)

145. Dobkin, D., Bartholomew, L., McDaniel, G, and DeDontney, J., Atmospheric Pressure Chemical Vapor Deposition of Tungsten Silicide, *J. Electrochem. Soc.*, 137-5, 1623- 1626 (May 1990)

146. Gross, M.E. and Schnoes, K.J., Chemical Vapor Deposition of Cobalt and Formation of Cobalt Disilicide, *Proc. 10th. Int. Conf. on CVD* (G. Cullen, Ed.), 759-765, Electrochem. Soc., Pennington, NJ 08534 (1987)

147. Paidassi, S. and Spitz, J., Study of Nb-Si Compounds Deposited by Chemical Vapor Deposition, *J. Less Common Met.* 61(2):213-220 (Oct. 1978)

148. Wang, M.S. and Spear, K.E., Experimental and Thermodynamic Investigations of the V-Si-H-Cl CVD System, *Proc. 9th. Int. Conf. on CVD* (M. Robinson et al, Eds.) 98-111, Electrochem. Soc., Pennington, NJ 08534 (1984)

149. Chu, S., Chu, T., Han, K., Liu, Y. and Mantravadi, M., Chemical Vapor Deposition of Cadmium Tellurium Films for Photovoltaic Devices, *Proc. 10th. Int. Conf. on CVD* (G. Cullen, Ed.) 982-989, Electrochem. Soc., Pennington, NJ 08534 (1987)

150. Ahlgren, W.L., Jensen, J.E. and Olson, G.L.,Laser-Assisted MOCVD of CdTe, *Proc. 11th. Int. Conf. on CVD* (K. Spear and G. Cullen, Eds.) 617-625, Electrochem. Soc., Pennington, NJ 08534 (1990)

151. Gentilman, R.L., Bibenedetto, B., Tustison, R. and Pappis, J., Chemical Vapor Deposition of Ceramics for Infrared Windows, in *Chemically Vapor Deposited Coatings*, (H. Pierson, Ed.) 47-53, The Amer. Ceram. Soc., Columbus OH 43214 (1981)

152. Collins, A.K. and Taylor, R.L., Optical Characterization of Polycrystalline ZnS Produced via Chemical Vapor Deposition, *Proc. 11th. Int. Conf. on CVD* (K. Spear and G. Cullen, Eds.) 626-633, Electrochem. Soc., Pennington, NJ 08534 (1990)

8

CVD IN ELECTRONIC APPLICATIONS

1.0 INTRODUCTION

The start of the solid-state electronic industry is generally recognized as 1947 when Bardeen, Brattain and Shockley of Bell Telephone Laboratories demonstrated the transistor function with alloyed germanium. The first silicon transistor was introduced in 1954 by Texas Instruments and, in 1956, Bell Laboratories produced the first diffused junction obtained by doping. The first-solid state transistor diodes and resistors had a single electrical function and were (and still are) known as discrete devices. In 1959, several components were placed on a single chip for the first time by Texas Instruments, inaugurating the era of integrated circuits (IC's) (1).

The major result of these developments was a drastic price reduction in all aspects of solid-state circuitry. The cost per unit of information (bit) has dropped by an estimated three orders of magnitude in the last twenty years.

This has been accompanied by a continuous decrease in size and today circuit integration has reached the point where up to a million components can be put on a single chip.

This dramatic progress is largely due to the development of thin film technologies such as evaporation, sputtering and chemical vapor deposition (CVD). Thin films play a critical role in the development of modern IC's and their processing comprises an increasingly high percentage of the production cost, estimated now up to 40% in some of the latest designs. In recent years, CVD has become the major thin film process and its share in the overall production of electronic and semiconductor components is continuously on the increase.

The semiconductor industry has grown very rapidly both in the US and abroad with a worldwide market estimated to reach $50 billion in 1990.

2.0 ELECTRONIC FUNCTIONS AND SYSTEMS

In order to obtain a clear view of the role of CVD in the technology of the electronic industry, it is important to have an understanding of the various electronic functions and systems.

2.1 Conductors, Semiconductors and Insulators

An electric current can be defined as a flow of electrons. In conductors such as metals, the attraction between the outer electrons and the nucleus of the atom is weak, the outer electrons can move readily and consequently metals are good conductors of electricity. In other materials, electrons are strongly bonded to the nucleus and are not free to move. Such materials are insulators (or dielectrics) (1,2). In semiconductors, as the name implies, the conductivity falls between those of conductors and insulators. Table 1 lists the characteristics of all three groups.

TABLE 1

Electrical Characteristics of Materials

Device	Typical Material	Electron Mobility	Resistivity ohm-cm
Conductor	Copper Silver Gold Tungsten Silicides	Free to move	10^{-5} to 10^{-6}
Semiconductor	Silicon Germanium III-V, II-VI compounds	Partially able to move	10^{-2} to 10^{9}
Insulator (dielectrics)	Alumina Silica Silicon nitride Glass	Bound to nucleus	10^{12} - 10^{22}

2.2 Categories of Electronic Devices

Electronic devices are either discrete devices or integrated circuits (IC's). Discrete devices perform a single function and include the following:

• Transistors, which control the current through a junction

of semiconductor materials by a voltage signal from an emitter electrode

• Resistors, which are passive devices limiting the flow of electrical current in proportion to the applied voltage and usually made of tantalum, nichrome, titanium or tungsten

• Thermistors, which are resistors made of semiconductor material having a resistance that varies rapidly and predictably with temperature

• Capacitors, which store an electric charge and consist of two conductors usually made of tantalum, titanium, nichrome, platinum or gold, separated by a dielectric, usually made of silica or silicon nitride

Integrated circuits (IC's) are circuits in which transistors, resistors, capacitors and their required connections are combined on a single chip of semiconductor material which is usually made of single crystal silicon.

Two major types of transistor commonly used in the electronic industry are essential to the operation of a computer: the bipolar transistor and the field-effect transistor or FET. The FET's can be divided in two subgroups: the n-channel metal-on-oxide FET (NMOS or MOSFET) and the complementary metal-on-oxide (CMOS). A typical FET design is shown schematically in Figure 1 (3). The sections produced by CVD are indicated in this diagram.

The trend in the production of IC's shows a gradual decrease in the relative importance of bipolar transistors and, conversely, an increase in digital MOS IC's which now dominate the market by a large margin. This trend is important with regard to CVD since CVD is used more extensively in the production of FET's than in that of bipolar transistors.

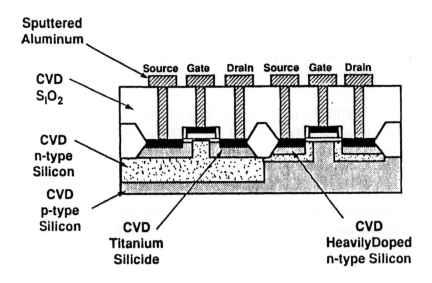

Figure 1. Typical Design of Field Effect Transistor (FET)

2.3 Device Miniaturization

The increase in the number of components that can be placed on an IC chip has been almost exponential in the last 30 years. This number now exceeds one million (ultra large scale integration or ULSI). The consequence is a considerable decrease in the minimum feature length, now routinely reaching submicron size.

With reduction in size comes reduction in intrinsic switching time and proportionally higher device speed. Digital IC's will soon be able to perform at rates reaching a gigabit/second (giga $= 1 \times 10^9$). This is accompanied by a corresponding drop in power consumption (3).

2.4 Strained Layer Superlattice (SLS)

Another concept for increasing speed is the strained layer superlattice (SLS), which consists of alternating layers of semiconductor materials with thickness < 100 Angstrom deposited by CVD. These materials have the same crystal structure but different lattice parameters. An example is an epitaxial layer of aluminum gallium arsenide ($Al_xGa_{1-x}As$) deposited on a layer of gallium arsenide. These two materials have the same cubic structure. The strain resulting from the differences in lattice parameters is accommodated elastically, i.e. there is no dislocation in the films, providing that these differences are not too large. Only very thin layers can accommodate such strains. If the thickness is too great , the strain can no longer be absorbed elastically and a dislocation occurs (see Chapter Two, Figure 11). The elastic strain alters the electronic band structure of the material, allowing conduction by light holes, thus greatly increasing the speed and reducing the power requirements. Practical devices known as high electron mobility transistors (HEMT) or ballistic transistors are in production by MOCVD.

2.5 Three-dimensional Structure

Another concept for high speed devices is based on the motion of electrons moving in a three-dimensional structure, across the thickness of the device in addition to parallel to the surface. The switching speed of a transistor is limited by the time required by an electron to move across the device (the gate in an FET or the base in a bipolar transistor). Moving through the thickness greatly reduces the switching time. This leads to improved analog systems such as radars with wavelengths in millimeters which can resolve much finer details than existing radars based on centimeter wavelength. Three-dimensional structures can also be used to build ULSI beyond the present million components per chip, using new material concepts such as silicon on sapphire (SOS), quantum well lasers, trenched transistors and capacitors, and buried layers. Such designs are only possible with MOCVD and molecular beam epitaxy (MBE) (see Appendix). MOCVD is presently more amenable to production than MBE and is used more extensively.

3.0 CVD IN ELECTRONIC TECHNOLOGY

CVD plays a major role in electronic technology. The fabrication of semiconductor chips is a complicated and lengthy process which, in new devices, may involve over one hundred steps. CVD has become an integral part of this process and is now used in the production of thin films of insulators, semiconductors and, more recently, conductors. The CVD of these films is estimated to average 20% or more of these processing steps which are summarized in Table 2.

An important consideration in the sequence of semiconductor devices fabrication is the so-called "thermal budget", a measure of both the CVD temperature and the time at that temperature for any given CVD operation. As a rule, the thermal budget becomes lower, the farther away a given step is from the original surface of the silicon wafer. This restriction is the result of the temperature limitations of the already deposited materials. For instance, doped phosphosilicate glasses used in planarization cannot be heated above 725°C which is their flow temperature. After a layer of aluminum is deposited, subsequent temperatures cannot exceed 380°C because spiking and the formation of hillocks would occur rapidly. The factor of time at a given temperature is just as important, as it will influence phenomena such as diffusion and dissolution. In the planning of a CVD process, the sequence of events and the thermal budget are essential considerations.

TABLE 2

Typical CVD Processing Steps in Chip Fabrication

PROCESS

System Process	Temp. °C	Applications
OXIDE		
Undoped oxide	250-450	Interlayer dielectric, masking, passivation (a)
Phosphosilicate glass	"	Reflow, doping source, passivation (a)
Borosilicate glass	"	Etch stop and doping source (a)
Two-step oxide	"	Intermetal dielectric (a)
Low-temp. oxide	"	Passivation (a)
SILICON		
Amorphous silicon	450-700	Gate interconnects (b)
Polysilicon	"	Gate interconnects and load resistors (b)
In situ phosphorus doping	"	Gate interconnects and load resistors (b)
TUNGSTEN		
Selective tungsten	300-600	Contact barrier, etch stop (a)

TABLE 2 (cont.)

Typical CVD Processing Steps in Chip Fabrication

PROCESS

System Process	Temp.°C	Applications
TUNGSTEN (cont.)		
Blanket tungsten	300-600	Metallization (a)
Amorphous silicon	"	Gate interconnects (a)
Polysilicon	"	Gate interconnects and load resistors (b)
NITRIDES		
Silane based	550-850	Interlayer dielectric (a)
Dichlorosilane based	"	Masking (a)
Amorphous silicon	"	Gate interconnects (b)
Polysilicon	"	Gate interconnects and load resistors (b)
HIGH TEMPERATURE OXIDE		
Stoichiometric oxide	650-850	Interlayer dielectric and masking (c)
Silicon-rich oxide	"	Semi-insulator (d)

TABLE 2 (cont.)

Typical CVD Processing Steps in Chip Fabrication

(a) Used in NMOS, CMOS and bipolar devices
(b) Used in NMOS and CMOS devices
(c) Used in NMOS devices
(d) Used in NMOS EPROM devices

4.0 THE CVD OF ELECTRICAL INSULATORS

Thin films of electrical insulators are an essential element in the design and fabrication of electronic components. The most widely used insulator materials are silicon dioxide, SiO_2, silicon nitride, Si_3N_4, and their modifications such as phosphorus-doped SiO_2 and silicon oxynitride. These materials are produced by CVD almost exclusively (4).

4.1 Silicon Dioxide (SiO_2)

The uses of CVD silicon dioxide films are numerous and include insulation between conductive layers, diffusion masks and ion-implantation masks for the diffusion of doped oxides, passivation against abrasion, scratches and the penetration of impurities and moisture. Indeed, SiO_2 has been called the pivotal material of IC's (5). Several CVD reactions are presently used in the production of SiO_2 films, each having somewhat different characteristics. These reactions are described in Chapter Seven, Section 5.5.

Some of the most significant developments in the CVD of SiO$_2$ include experiments in plasma CVD at 350°C via electron cyclotron resonance (ECR) (described in Chapter Three) to gain improved control of the deposition rate and obtain a quality equivalent to that of the thermally grown oxide. The use of deposition from diacetoxyditertiarybutoxy silane at 450°C has also been shown to significantly improve the film properties of SiO$_2$(6).

A recent competitor to the CVD in the planarization of silicon dioxide is the sol-gel process, where tetraethyl orthosilicate is used in the form of spin-on-glass (SOG) film.

This technique produces films with good dielectric properties and resistance to cracking. Gas-phase precipitation, which sometimes is a problem with CVD, is eliminated (7). Over the next few years we will learn how serious a competitor sol-gel can be.

4.2 Silicon Nitride

Silicon nitride (Si$_3$N$_4$) is a very hard and scratch-resistant electrical insulator. In contrast with SiO$_2$, it is an excellent diffusion barrier, especially for sodium and water which are major sources of corrosion and instability in microelectronic devices. As a result, silicon nitride can perform two functions simultaneously: passivation and provision of a diffusion barrier. It is used with increasing frequency in the fabrication of IC's in such areas as oxide insulation masking (to be removed during subsequent processing), local oxidation of silicon (LOCO), dielectric between two layers of polysilicon for capacitors in analog cells and in NMOS.

Silicon nitride is produced almost exclusively by CVD. The reactions are described in Chapter Seven, Section 4.5. Silicon nitride is normally produced by plasma CVD. This process does not require high temperature, which is a desirable feature particularly in passivation, as it avoids reactions between the silica and the metallic conductor.

5.0 THE CVD OF SEMICONDUCTORS

5.1 Silicon

The electronic properties of silicon and other semiconductor materials are shown in Table 3.

TABLE 3

Electronic Properties of Semiconductor Materials

Property	CVD diamond	Si	GaAs	βSiC
Dielectric constant (198Hz)	5.5	11.8	10.9	9.7
Dielectric strength, Mv/cm	10	0.3-0.5	0.6	4.0
Bandgap (eV)	5.40	1.12	1.43	2.3-3.0
Saturated electron velocity (10^7cm/s)	2.7	1.0	2.0	2.5
Carrier mobilities $(\text{cm}^2/\text{V.s})$				
electron (n)	-	1350-1500	8500	400-1000
positive hole (p)	-	480	400	50

TABLE 3 (cont.)

Electronic Properties of Semiconductor Materials

Property	CVD diamond	Si	GaAs	βSiC
Semiconductor temperature limit (°C)	500	150	250	>1000

Note: The CVD of diamond is still considered experimental

In many ways, silicon is the ideal semiconductor material. It is readily available with a very high degree of purity and relatively low cost. Most IC's are made from silicon and this is likely to remain so for some time (6). CVD is used in silicon production in two major areas: 1) in the production of ultra-pure silicon from which single-crystal ingots are made, and 2) in the preparation of epitaxial and polycrystalline films.

Single Crystal Processing: The silicon chips are fabricated by slicing a cylindrical ingot of single crystal silicon into thin wafers of up to 200 mm in diameter. Each wafer is then divided into approximately 100 chips. The performance of IC's depends to a great degree on the integrity and purity of these chips. Consequently the production of the single crystal silicon ingot is most critical and requires very close control of impurities and fabrication process. CVD plays an important role in this process since it is the basic method for preparation of electronic grade silicon (EGS). The single crystal ingot is obtained from the melted ESG by a crystal pulling operation (6) (see Chapter Six, Section 5.2).

Epitaxial Silicon: Epitaxy is a term that denotes the growth of a thin crystalline film on a crystalline substrate. When the epitaxial film is of the same material as the substrate (for instance silicon on silicon), the process is known as homoepitaxy. When film and substrate are of different materials, it is heteroepitaxy. Epitaxy is described in Chapter Two, Section 4. CVD epitaxial silicon films were developed originally to improve the performance of discrete bipolar transistors. They are now used extensively in FET's, dynamic random access memory devices (RAM) and other IC designs. In short, epitaxial silicon films are an indispensable part of every microelectronic system (8).

The CVD reactions used to produce epitaxial silicon are described in Chapter Six, Section 5.2. Originally, atmospheric-pressure CVD was used, but it is gradually being replaced by low-pressure CVD in spite of higher equipment cost and complexity. Low-pressure CVD appears to yield a better film with less autodoping and pattern shift. To provide the necessary semiconductor properties, epitaxial films are normally doped by incorporating boron, arsenic or phosphorus in the silicon structure by the addition of gases such as arsine, diborane and phosphine during the deposition.

Phosphorus as a dopant reduces internal stresses, increases the moisture resistance and is used as an alkali getter (mostly of sodium). Boron as a dopant decreases the etch rate and improves step coverage (5,7). The CVD reactions are described in Chapter Three, Section 6 and Chapter Four, Section 8. Doping during CVD is gradually replacing doping by diffusion or by ion implantation.

Polysilicon: "Polysilicon" and "poly-Si" are contractions of polycrystalline silicon (in contrast with the single-crystal epitaxial silicon). Like epitaxial silicon, polysilicon is also used extensively in the fabrication of IC's and is deposited by CVD. Examples of polysilicon films are:

• gate electrodes in MOS devices for high value resistors to insure good ohmic contact to crystalline silicon

• diffusion sources to form shallow junctions

• emitters in bipolar technology

Polysilicon is doped in the same manner as epitaxial silicon.

5.2 Diamond

Another semiconductor material of great interest is CVD diamond, although the technology is still in the experimental stage (see Chapter Six, Section 4). Pure single crystal diamond is one of the best electrical insulators. However, the presence of impurities can drastically alter its electronic state and the inclusion of even very small quantities of sp^2 carbon will render the material useless for electronic applications. By suitable doping, diamond can be made into a semiconductor (9).

With excellent semiconductor properties, diamond has a very good potential of becoming as useful semiconductor material (10). It has the widest bandgap of any semiconductor and an upper limiting semiconductor temperature of approximately 500°C. In comparison, the upper limit of silicon is 150°C and that of GaAs is 250°C.

Diamond has a high electron carrier mobility exceeded in the p-type only by germanium and, in the n-type, only by gallium arsenide. The saturated carrier velocity, that is the velocity at which electrons move in high electric fields, is higher than for silicon, gallium arsenide or silicon carbide. Unlike other semiconductors, semiconductor diamond retains its high saturated current velocity at high temperatures.

The semiconductor properties of a material can be summarized in two figures of merit:

The Johnson figure of merit, based on saturated carrier velocity and dielectric strength (product of power x frequency squared x impedance) which predicts the suitability of a material for high power applications. It is normalized with the value of one given to silicon. As shown in Table 4 below, diamond is clearly the preferred material on this basis.

The Keyes figure of merit, based on thermal conductivity as an additional factor (see following section) which predicts the suitability of a semiconductor for dense logic circuit applications.

Again, diamond is far superior to other materials.

TABLE 4

Figures of Merit for Semiconductor Materials

	Si	GaAs	InP	Diamond
Johnson figure of merit (high power applications)	1	7	16	1138
Keyes figure of merit (dense logic circuits)	1	0.5	6	32

A number of semiconductor devices based on CVD diamond are being developed (11). One is a Schottky power diode capable of operating at 500°C, which has shown good rectification. Such a design would be ideal for high-voltage transistors with no need for voltage conversion from line to input. This would considerably reduce the size of the power supply.

Another CVD diamond semiconductor application leads to a series of field effect transistors (MOSFET and MESFET) which are proving superior to silicon devices and are characterized by high power handling capacity, low saturation resistance and excellent high frequency performance. It is well recognized that CVD diamond would be the ideal material for semiconductor applications particularly in high power and high frequency systems, such as microwave, as well as in harsh environments such as in internal combustion and jet engines. However many problems must be solved before practical, reliable and cost-effective diamond semiconductor devices can be produced. Yet, the prospects are good, particularly if epitaxial single crystal diamond can be effectively deposited.

5.3 III-V and II-VI Semiconductors

The III-V and II-VI compounds refer to combination of elements that have two, three, five or six valence electrons. They have semiconductor properties and are all produced by CVD either experimentally or in production. The CVD of these materials is reviewed in Chapter Six, Section 6. Many of their applications are found in optoelectronics where they are used instead of silicon since they have excellent optoelectronic properties. Silicon is not a generally satisfactory optical material since it emits and absorbs radiation in the range of heat instead of light.

The best known and most widely used of the II-V and II-VI

materials is gallium arsenide which is emerging as an important complement to silicon. Compared to silicon, it has the following characteristics:

• superior speed and power capability

• higher operating temperature

• capability of emitting light and ability to combine the processing of both photons and electrons on a single chip

• easily alloyed with other III-V compounds

• high frequency capability (RF and MW)

For all its advantages, gallium arsenide has yet to be used on any large scale, at least outside optoelectronic applications. The reasons are cost (over ten times that of silicon), small wafer size, low thermal conductivity (1/3 that of silicon) and low strength. However, a new process based on CVD combines its advantages with those of silicon. With the CVD epitaxial process, it is possible to deposit a gallium arsenide epilayer on a silicon substrate. The resulting composite combines the mechanical and thermal properties of silicon with the photonic capabilities and fast electronics of gallium arsenide (12,13).

Other gallium-based III-V compounds are also produced by CVD, either experimentally or in production. The major ones are listed in Table 5.

The many possible combinations of gallium compounds allows the tailoring of electronic and optoelectronic properties to suit specific applications. Of particular importance is the control of the stoichimetric ratio(s) of the element involved. This is achieved by the proper handling of the MOCVD reactions. Being able to tailor the bandgap imparts great flexibility in the design of transistors and optoelectronic devices.

TABLE 5

Optoelectronic Properties of Gallium Compounds

Material	E max. (eV)	Recombination process	Emission
$GaAs_{1-x}P_x$	2.0	Direct gap	Red
$Ga_{1-x}Al_xAs$	1.9	Direct gap	Red
GaP	2.3	Indirect gap	Red, yellow, green
$Ga_xIn_{1-x}P$	2.2	Direct gap	Yellow
$Ga_{1-x}Al_xP$		Pair emission	Green

The bandgap is defined as the energy difference between the top of the valence band of an atom and the bottom of the conduction band (14). The bandgap values of several semiconductor materials are shown in Chapter Two, Figure 12. The bandgap of GaAs is slightly greater than that of silicon and that of AlAs is considerably greater. By alloying GaAs with Al to form $Ga_xAl_{1-x}As$ or by partially substituting Al for Ga, a wide range of bandgaps can be produced (Figure 2) (15). Indium can also be substituted for gallium, phosphorus for arsenic, and many other combinations are possible by the careful control of the CVD process. II-VI materials include the compounds of arsenic, cadmium and mercury with sulfur, selenium and tellurium (see Chapter Nine, Section 2.0).

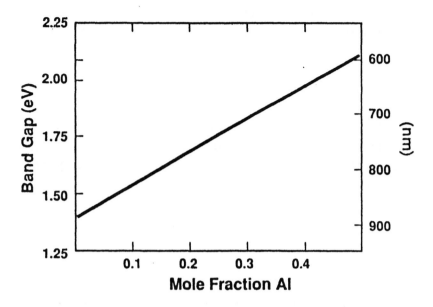

Figure 2. Bandgap of Aluminum Gallium Arsenide as a Function of Composition at 300K

Most of these compounds were originally prepared by liquid-phase epitaxy. That process is now largely replaced by MOCVD, particularly in the case of gallium arsenide, gallium arsenic phosphide and gallium aluminum phosphide.

6.0 THE CVD OF ELECTRICAL CONDUCTORS

Aluminum has long been the metal of choice for electrical conductors. It has a low vaporizing temperature and is well suited to deposition by evaporation. It is, however, susceptible to electromigration, i.e. the movement of metal from one part of the line to another. This phenomenon occurs when a very high current density (greater than 10^6 Amp/cm^2) is used, which is a common occurrence in new circuit designs. Once electromigration is started, it can only intensify. To mitigate the problem, alloys of aluminum and copper or aluminum and silicon are used. These alloys are usually deposited by bias sputtering. However, they offer only a temporary solution as electromigration will still occur as greater densities of circuit elements are introduced.

Another problem associated with aluminum and its alloys is their reaction with silicon which leads to leakage and shorting. A solution to these problems is found in the use of refractory metals and silicides (16).

6.1 Refractory Metals

The major refractory metal of interest in microelectronic devices is tungsten which is beginning to be used extensively in VLSI interconnections. Tungsten has excellent electromigration resistance and relatively low resistivity (5-6 μohm-cm). Originally applied by sputtering, it is now produced mostly by CVD, generally using the hydrogen reduction of tungsten hexafluoride. A cold-wall reactor is normally used with lamp heating of the substrate. This reaction and other tungsten deposition reactions are described in Chapter Five, Section 19.

New high-density IC designs may have over one million contact openings connecting the first layer of metal to the source/drain. Every one of these holes must be completly filled with metal either by a blanket deposit followed by etching, or by selective deposition. This was previously achieved with sputtered aluminum which is now being replaced by CVD tungsten.

The advantages of CVD tungsten can be summarized as follows (17):

• Deposition can be selective or non-selective

- No patterning is necessary for selective deposition

- Non-selective films are easily patternable by wet or dry etching technology

- Blanket deposition has excellent conformal coverage

- Films can be deposited over a wide range of temperature

- Electromigration is far less than that of aluminum

- A good barrier to Al-Si interaction is provided up to 450°C for 30 min.

- Films have low stress on Si and other substrates

- Films have low contact resistance, especially if silicide layers are deposited between W and Si

- Films do not induce excessively high leakage currents in silicon junctions, provided that junction depths are greater than about 2500 Angstroms

Certain problems remain to be solved before tungsten can be adopted on a large scale for metallization. A major problem is the erosion of silicon during deposition which occurs according to the following reaction:

$$2WF_6 + 3Si \longrightarrow 2W + 3SiF_4$$

This reaction proceeds at a much faster rate than the hydrogen reduction of WF_6. The result is erosion of the silicon substrate causing encroachment and tunnel defects. The use of a different precursor such as tungsten carbonyl, $W(CO)_6$, may solve this problem.

CVD tungsten is presently limited mostly to multilevel via fill applications. However, once the problem of silicon erosion is solved, a considerable increase in the use of tungsten for metallization will undoubtedly occur.

Molybdenum is another refractory metal with low resistivity (5-7 μohm-cm) now under investigation for metallization of IC's. It is usually deposited by the decomposition of the carbonyl, $Mo(CO)_6$, or by the hydrogen reduction of the halides ($MoCl_5$ or MoF_6). These reactions are described in Chapter Five, Section 10.

6.2 Silicides

Metal silicides are used increasingly as conductive thin films and are rapidly replacing metals in a number of applications. They are also replacing polycrystalline silicon conductors in VLSI's because of their low resistivity (1/10th. that of silicon), their much higher thermal stability and their excellent diffusion barrier characteristics (12,18).

Of particular interest are the refractory metal silicides such as $MoSi_2$, $TiSi_2$, WSi_2, $TaSi_2$ and $CoSi_2$, which are especially suit-able for gate metallization, and the silicides of the platinum group metals, which are suitable for making contacts.

The most widely used silicides at present are WSi_2 and $TaSi_2$. WSi_2 is deposited almost exclusively by LPCVD. However, in the case of CVD $TaSi_2$, the film suffers from undesirable growth morphology and further research is needed. It is now mostly deposited by sputtering.

The "salicide" concept (self aligned silicide) is now used extensively in MOSFET devices. It is based on $TiSi_2$ and $CoSi_2$, both of which are deposited by CVD (19,20). The CVD reactions for the deposition of silicides are described in Chapter Seven, Section 6.

The trend in CVD metallization is toward the extensive use of refractory metals and their silicides in multilayered metallization designs typically consisting of metal silicide contacts, refractory metal barriers and an aluminum alloy as the principal interconnect metal. Other CVD metals are also actively considered for use as conductors such as chromium, copper, molybdenum, platinum, rhodium and ruthenium.

7.0 THE CVD OF SUBSTRATES

The density of integrated circuits (ICs) used in microwave, millimeter-wave and photonic semiconductor devices is presently limited by the large amount of heat generated by the extremely close packing of the electronic components on the chip and the very high frequencies and power density levels. To remove this heat, it is necessary to use hybrid circuits and bulky heat- dissipation devices or complicated and expensive refrigeration. Conventional substrate materials such as alumina are becoming inadequate and new heat-sink materials are necessary.

Likewise, advances in thin-film deposition technology have led to the production of laser diodes with extremely high intensity of light output, an output that is further augmented when the diodes are placed in arrays. Here again, the limiting factor in the performance of the devices is the ability to dissipate the heat in the package.

This problem can be solved, or at least alleviated, by using an electrically insulative substrate with high thermal conductivity. The few materials that have these characteristics are listed in Table 6.

TABLE 6

Properties of Substrate Materials

	CVD Diamond	AlN	BeO	Al_2O_3
Resistivity ohm-cm	10^{12}	10^{14}	10^{14}	10^{14}
Density g/cm³	3.51	3.26	3.01	3.97
Coef. of thermal exp. ppm/°C (0-400°C) (silicon is 3.8)	2.8	4.6	7.6	7
Thermal conductivity W/m K	1800	185	240	29

Diamond deposited by CVD offers the best potential, but so far, it has only been used on an experimental basis and there are still problems in thickness control and uniformity. Beryllia is an excellent heat-sink material which is presently widely used but is being phased out because it presents acute safety problems (21).

CVD aluminum nitride is now the material of choice and is being actively considered for a number of applications (22). The corresponding CVD reactions are described in Chapter Seven, Section 4.1.

8.0 THE CVD OF SUPERCONDUCTORS

The recent discovery of high temperature superconductivity in mixed oxides, such as the lanthanum-barium-copper oxide complexes, has created a great deal of interest in these materials. Superconductivity, that is, the absence of any resistance to the flow of electric current, is now an established fact at temperatures above the temperature of liquid nitrogen (77K).

Many problems remain in the development of practical processes for these materials and commercialization is not likely to occur until these problems are solved (23). Among the several processing techniques now used, CVD appears one of the most successful.

Table 7 lists the principal copper oxide complexes now under development and their properties (24). The CVD reactions used to deposit these materials are described in Chapter Seven, Section 5.12.

It appears now that the most likely applications of these superconductors to reach the practical stage will be in the form of coatings, mostly for semiconductor and electronic and related applications.

For larger current-carrying applications, a superconductor coating over a metallic conductor such as copper may also become a practical design because of its advantage over a monolithic superconductor wire. It is able to handle current excursions and has better mechanical properties.

The metal alloy, niobium germanium (Nb_3Ge), is another superconductor with a much lower transition temperature (20K) with well-established characteristics and good strength. It is deposited by CVD on an experimental basis by the reaction described in Chapter Five, Section 20.5.

TABLE 7

Properties of Superconducting Materials

Compound	Transition temperature K	Current capacity A/cm^2
$YBa_2Cu_3O_7$	93	10000
$Bi_2Sr_2CaCu_2O_x$	110	2000
$Tl_2Ca_2Ba_2Cu_3O_x$	125	1000

Applications of CVD superconductors now being considered include the following:

• Josephson junctions for passive microwave devices, satellite communication systems and computer logic gates

• Magnetic field detectors (superconducting quantum interference devices or SQUIDS)

• Magnet wires of $YBa_2Cu_3O_7$ for motor generator and energy storage systems

• Superconducting short dipole antenna materials with a resonant frequency of 550 MHz (25)

REFERENCES

1. Sze, S.M., *Semiconductor Physics and Technology*, John Wiley & Sons, New York (1985)

2. Murray, R.L. and Cobb, G.C., *Physics, Concepts and Consequences*, Prentice Hall Inc., Englewood Cliffs, NJ (1970)

3. Chaudhari, P., Electronic and Magnetic Materials, *Scientific American*, 137-146 (Oct. 1986)

4. Blum, J. M., CVD of Dielectrics, a Review, *Proc. 10th. International Conf. on CVD*, (G. Cullen, Ed.), pp. 476-489, Electrochemical Soc., Pennington, NJ 08534 (1987)

5. Pramanik, D., CVD Dielectric Films for VLSI, *Semiconductor International*, 94-99 (June 1988)

6. Wang, P. and Bracken, R.C., Electronic Applications of CVD, *Proc. 10th. Int. Conf. on CVD*, (G. Cullen, Ed.) pp. 755-787, Electrochem. Soc., Pennington, NJ 08534 (1987)

7. Elliott, J.K., Current Trends in VLSI Materials Part II, *Semiconductor International*, 150-153 (April 1988)

8. Goulding, M.R. and Borland, J., Low vs. High Temperature Epitaxial Growth, *Semiconductor International*, 90-97 (May 1988)

9. Fujimori, N., Diamond Semiconductors, *New Diamond* (Japan) 2-2:10-15 (1988)

10. Spear, K., Diamond, Ceramic Coating of the Future, *J. Am. Cer. Soc.*, 72(2):171-91 (1989)

11. Orr, R., The Impact of Thin Film Diamond on Advanced Engineering Systems, *Proc. of Conf. on High Performance Inorganic Coatings*, G.A.M.I., Gorham, ME 04038 (1988)

12. Elliott, J.K., Current Trends in VLSI Materials, Part I, *Semiconductor International*, 46-51. (March 1988)

13. Chen, H.Z., Paslaski, J., Yariv, A. and Morkoc, H., Combine the Strength of Silicon with the Power of GaAs, *Research and Development*, 61-66 (Jan. 1988)

14. Brodsky, M.H., Progress in Gallium Arsenide Semiconductors, *Scientific American*, 68-75 (Feb. 1990)

15. Wolfson, R.G., Application Specific Electronic Materials by Ion Implantation and MOCVD, *Proc. of Conf. on High Performance Inorganic Coatings*, Monterey, CA, G.A.M.I., Gorham, ME 04038 (1988)

16. Green, M.L. and Levy, R.A., The CVD of Metals for Integrated Circuit Applications, *J. of Metals*, 63-71 (June 1985)

17. Green, M.L., The Current Status of CVD Tungsten as an Integrated Circuit Metallization, *Proc. 10th. Int. Conf. on CVD* (G. Cullen, Ed.) 603-613, Electrochem. Soc., Pennington, NJ 08534 (1987)

18. Gupta, S., Song, J.S. and Ramachandran, V., Materials for Contacts, Barriers and Interconnects, *Semiconductor International*, 80-87, (Oct. 1989)

19. Peauleau, Y., CVD of Refractory Metals and Refractory Metal Silicides, *Proc. 10th. Int. Conf. on CVD*, (G. Cullen, Ed.), pp. 685-699, Electrochem. Soc., (1987)

20. Murarka, S.P., *Silicides for VLSI Applications*, Academic Press, Boston MA (1983)

21. Kerney, K.M., Aluminum Nitride Expands Packaging Options, *Semiconductor International*, 32, (Dec. 1988)

22. Susuki, M. and Tanji, H., CVD of Polycrystalline Aluminum Nitride, *Proc. 10th. Int. Conf. on CVD*. (G. Cullen, Ed.) pp. 1089-1097, Electrochem. Soc., Pennington, NJ 08534 (1987)

23. Wilson, S., The Use of Thin Films in the Manufacture of Semiconductor IC's, *Proc. of Conf. on High Performance Inorganic Coatings,* Monterey, CA (1988), G.A.M.I., Gorham, ME 04038

24. Studt, T., Challenges and Opportunities in the New Age of Materials, *Research and Development*, 56-64, (Nov. 1990)

25. *Design News,* p.24, May 8, 1989

9

CVD IN OPTOELECTRONICS AND OTHER ELECTRONICALLY RELATED APPLICATIONS

1.0 CVD IN OPTOELECTRONICS

Optoelectronics is a branch of science and engineering which combines optics and electronics, in which CVD now plays a major part. Optoelectronics deals with optical wavelengths from 0.20 micron (ultraviolet) to 3 microns (near infrared) as shown in Figure 1. As can be expected, the properties of optoelectronic materials are a combination of electrical and semiconductor properties (electron action) with optical properties such as transmission, reflection and absorption (phonon action).

Optoelectronics is a new and fast growing industry with applications in many areas from television to watches to laser communications. Many applications in optoelectronics were developed in Japan and that country dominates the industry at the present time. CVD is used extensively in the fabrication of optoelectronic devices, both experimentally and in production.

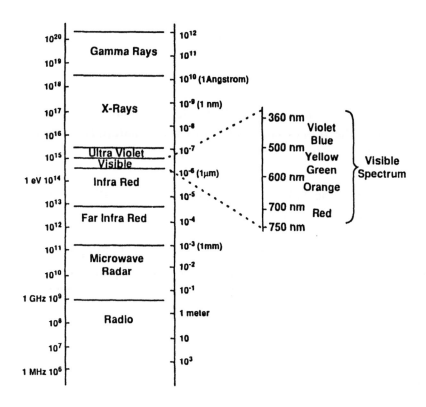

Figure 1. Electromagnetic Spectrum

2.0 OPTOELECTRONIC MATERIALS

Silicon is not generally considered a satisfactory material for optoelectronics since it emits and absorbs heat instead of light. However, this may change as recent experiments have shown that a red light can be produced by shining an unfocused green laser beam on a specially prepared ultrathin crystal silicon slice (1). CVD may prove useful in preparing such a material.

The III-V and II-VI semiconductor compounds have excellent optical properties and are the most important group of optoelectronic materials. They are all produced by CVD for electronic and optoelectronic applications. Their properties and CVD reactions are reviewed in Chapter Six, Section 7 and Chapter Eight, Section 5.3.

As stated in Chapter Eight, it is possible to tailor the bandgap by the proper combination of these materials, to suit any given application (see Chapter Eight, Figures 3 and 4).

Among the wide variety of combinations possible with these materials, the most common are the III-V compounds of gallium such as gallium arsenide, gallium arsenic phosphide, gallium aluminum arsenide, gallium phosphide, gallium indium phosphide, gallium indium arsenide and gallium aluminum phosphide. Other common compounds are indium arsenide (InAs), indium phosphide (InP), boron phosphide (BP), indium arsenide antimonide (InAsSb), aluminum arsenide (AlAs) and others.

Commonly used II-VI compounds include zinc sulfide, zinc selenide, zinc telluride, cadmium sulfide, cadmium telluride and mercury cadmium telluride. These materials are not as widely used as the III-V compounds, one reason being that it is difficult to achieve p-type doping. The major applications of the various II-VI compounds are found in photovoltaics and electroluminescent displays.

Other optoelectronic materials made by CVD include the ferroelectric compounds such as $Pb(Zr,Ti)O_3$ (PZT), PLZT which is a solid solution of $PbZrO_3$, $PbTiO_3$, La_2O_3, $KNbO_3$, $LiNbO_3$ and $BaTiO_3$ (2). These mixed oxides have ferroelectric properties (see Section 5.0 below) and their CVD reactions are reviewed in Chapter Seven, Section 5.11 (3).

3.0 OPTOELECTRONIC CVD APPLICATIONS

The optoelectronic compounds mentioned above are beginning to be used on a very large scale in numerous consumer applications. For instance, CVD gallium arsenide is a key component in the production of compact disk players, laser printers, radar detectors, cellular telephones, direct broadcast television and many others applications.

Recent optoelectronic devices and their applications, in which CVD plays a major part, are listed in Table 1 (4).

TABLE 1

Optoelectronic Devices and Applications

Device	Applications
Photodiode (LED)	Cameras, strobes, illuminators, remote controls, IR sensors
Pin photodiode	Fiber-optic communications, fiber links
Laser detector	Digital audio disks, video disks, laser-beam position sensors, distance sensors
Optic receiver	Optic remote controllers

TABLE 1 (cont.)

Optoelectronic Devices and Applications

Photodiode array (LED)	Photodiode chips can be arrayed monolithically or non-monolithically to any element
Photo transmitters	Optical switches, strobes, toys
IR-emitting diodes (GaAlAs)	Optical switches, encoders, photo IC sensors
IR-emitting diodes (GaAs)	Remote controls, optical switches, optoisolators, choppers, pattern recognition devices
Photo interrupter	Highly precise position sensors, non-contact switches
Reflective sensor	Tape-end sensors, liquid-level sensors
Photocoupler	Isolators, impedance converters, noise suppressors

Some of the most important optoelectronic devices are described below.

3.1 Light Emitting Diodes (LED)

Light emitters, such as the carbon arc or the incandescent lamp, are well-known but those of interest here are the light-emitting diode (LED) and the laser.

In a light-emitting diode of appropriate semiconductor material, light is emitted when electrons are made to flow across the p-n junction (Figure 2). Electrons must be energized with a certain voltage in order to cross the p-n junction. After crossing, they return to their normal state and, in so doing, produce light in a very narrow wavelength.

A variety of colors such as green, amber and red (and infrared) can be obtained with different semiconductor materials without the need for a filter (see Chapter Eight, Table 5). An LED (or photodiode) device may consist of multiple diodes in an array operating in the reverse-bias mode. Patterns of light showing symbols, letters or numbers can thus be produced with different colors obtained by doping the semiconductor material by CVD or ion implantation.

LED materials include gallium arsenic phosphide, gallium aluminum arsenide, gallium phosphide, gallium indium phosphide and gallium aluminum phosphide. The preferred deposition process is MOCVD, which permits very exacting control of the epitaxial growth and purity.

Typical applications of LED's include watches, clocks, scales, calculators, computers, optical transmission devices and many others.

3.2 Light Detectors

Light detectors fall into two categories: photoconductors and photodetectors. Photoconductors are devices whose resistance decreases upon exposure to light. Cadmium sulfide (CdS) and cadmium selenide (CdSe) are the most commonly used photoconductor materials in the visible spectrum. They are still mostly produced by sputtering but CVD is used increasingly (see Chapter Seven, Section 7.0).

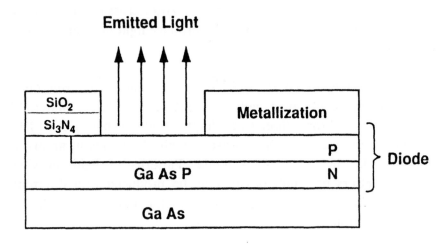

Figure 2. Schematic Diagram of Typical Light Emitting Diode

Photodetectors operate by carrier transport across a semiconductor junction. A wide variety of these photodiodes are available such as Schottky diodes, phototransistors and avalanche photodetectors. Typical photodetector materials are gallium arsenic phosphide and gallium phosphide, which are now produced by MOCVD.

3.3 Semiconductor Lasers

The light produced by a laser has a much more narrow wavelength than the light of an LED or other light sources (Figure 3). In addition, laser light is coherent, i.e. the photons travel in parallel paths from the source. The design of semiconductor lasers is similar to that of LED's and consists of a p-n junction between two different direct bandgap semiconductors with light-emitting capability.

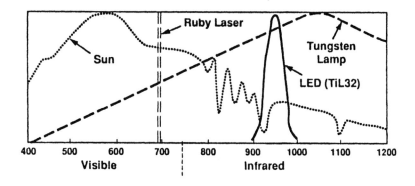

Figure 3. Radiation Wavelengths of the Sun, an LED, a Ruby Laser, and a Tungsten Lamp.

Until recently, most semiconductor lasers were made by liquid epitaxy but new design requirements call for much tighter specifications which can only be achieved by MOCVD or molecular beam epitaxy (MBE). Some of these new laser designs are based on the quantum-well principle in which a very thin layer of a semiconductor with a high band gap such as gallium aluminum arsenide is sandwiched between two films of GaAs which has a smaller band gap. This design can increase by tenfold the amount of generated light over that emitted by a standard semiconductor laser.

Since only a very low current is required, it is possible to build an array of such lasers on a single chip. This may lead the way to optical logic circuits for computers. A similar design uses active layers of gallium indium arsenic phosphide, 600 Å thick, sandwiched between InP buried layers for integrated digital networks. Another design uses four alternate layers of indium gallium arsenide and indium gallium arsenic phosphide, each 60-70 Å thick which are deposited on a diffraction lattice substrate for lasers. These materials are all produced by MOCVD (5,6).

3.4 Trends in CVD Optoelectronic Applications

MOCVD is presently the method of choice for fabricating III-V and II-VI semiconductor devices. Its competition is molecular beam epitaxy (MBE) which is described in the Appendix. Both MOCVD and MBE produce high-quality materials with similar purity, mobility and photo-luminescence properties, but MOCVD is more suitable for large-scale production since MBE requires very high vacuum and has low deposition rates.

In addition, MOCVD is better for phosphorous compounds, for aluminum gallium arsenide and for alloys with two Group V elements (As, Sb, Bi). It also seems to yield better optoelectronic and heterojunction bipolar transistor structures (7). On the other hand MBE can achieve better film growth control and more abrupt interfaces.

The two processes, MOCVD and MBE, are becoming closer in concept. For instance, MOCVD is using techniques developed for MBE such as in situ characterization monitoring, load lock and lower pressure levels and MBE is now using chemical sources such as organometallics which are typical of CVD.

Epitaxial deposition of gallium arsenide and other III- V compounds is fast becoming a preferred system in the design of new devices such as HEMT (high electron mobility transistor), where sequential layers are required that may be as thin as 20 Å. In such applications, epitaxial GaAs can only be obtained by MOCVD or by MBE.

4.0 CVD IN PHOTOVOLTAICS

A photovoltaic material generates a voltage when it is exposed to light and photovoltaics can be considered as a specialized area of optoelectronics. The principle has been known for many decades but it became a reality only in 1958 when an array of photovoltaic cells provided power for a space vehicle.

Until recently, photovoltaic devices were restricted mostly to space applications due to their very high cost. In the last few years, however, cost has been reduced by more than an order of magnitude and the spectrum of applications is now much broader. The cost, while still high (slightly less than $10/watt of generating capacity in 1990) is declining, as technology, and particularly CVD technology, is improving.

Besides space-power generation, the most publicized applications are in experimental grid power plants intended for eventual competition with classical power sources for large-scale generation of electricity.

Promising and profitable applications are found when solar cells are included in many consumer products such as watches, calculators and many others. Most of these products are produced in Japan.

The photovoltaic industry is continuously expanding. Its development is largely the result of advances in materials and thin-film technology. The deposition techniques, which were at first based mostly on sputtering and evaporation, are now increasingly relying on CVD.

4.1 Photovoltaic Principle and Operation

A photovoltaic cell is basically a semiconductor diode consisting of a junction similar to the junction of a transistor described in the previous chapter. An electrical potential is formed by n-type doping on one side and p-type on the other. Under the impact of light (photons) such as in sunlight, electrons move from the p side, across the junction to the n side, and, through electrical contacts, can be drawn as a usable current (Figure 4).

A major limitation of photovoltaic cells is their low efficiency. Sunlight consists of photons with a large range of frequencies, and consequently, an equally large range of energies. However, a given photovoltaic material responds only to the photons which have sufficient energy to lift an electron from the valence band to the conductive band across the bandgap. Photons with insufficient energy merely pass through without any effect or with detrimental heating of the cell.

Photovoltaic materials are needed that respond to the widest possible range of photon frequency, such as amorphous silicon, gallium arsenide and other III-V compounds. They have the characteristic of absorbing the long wavelengths of the sunlight spectrum and are efficient photovoltaic materials.

The thickness of a photovoltaic cell is chosen on the basis of its ability to absorb sunlight, which in turn depends on the bandgap and absorption coefficient of the semiconductor. For instance, it takes 50 microns of crystalline silicon to absorb the same amount of sunlight as one micron of amorphous silicon and 0.1 micron of copper-indium diselenide. Thin-film processes and CVD in particular are an excellent choice for producing such very thin films (8).

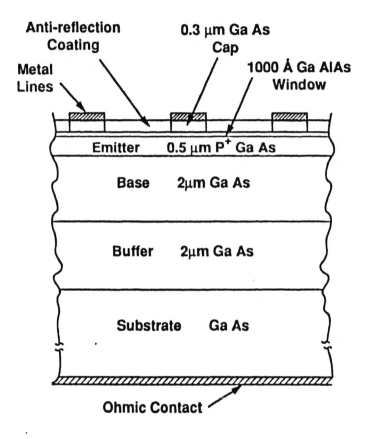

Figure 4. Typical Photovoltaic (Solar) Cell

4.2 Photovoltaic Material and Processing

Many materials with photovoltaic capability are known but only the following have been developed to any extent.

Single Crystal Silicon: Silicon is still the dominant material in photovoltaics. It has good efficiency which is 25% in theory and 15% in actual practice. Silicon photovoltaic devices are made from wafers sliced from single crystal silicon ingots, produced in part by CVD (see Chapter Eight, Section 5.1). However silicon wafers are still costly, their size is limited and they cannot be sliced to thicknesses less than 150 microns. One crystalline silicon wafer yields only one solar cell, which has an output of only one watt. This means that such cells will always be expensive and can only be used where their high efficiency is essential and cost is not a major factor such as in a spacecraft.

Amorphous Silicon (a-Si): Amorphous silicon is considered a promising new material (9, 10). As mentioned above, only a very thin coating is necessary since the amorphous structure is much better at absorbing sunlight than is the crystalline material. The most common process to produce a-Si is the decomposition of silane by plasma CVD (see Chapter Six, Section 5.2) (11).

This method is still slow and not entirely free of deficiencies. Thicknesses of a few microns can be deposited and, in contrast with single crystal silicon, relatively large areas can be covered. The cost can conceivably be reduced to levels low enough that solar cells of amorphous silicon could be considered for large-scale power generation.

The drawbacks of a-Si are: an efficiency that is still low and a lack of stability, as the material degrades and loses efficiency when first exposed to sunlight (Staebler-Wronski effect). However, the efficiency stabilizes in time.

A thin film of tin oxide with a rough texture, produced by MOCVD from tetramethyl tin, $(CH_3)_4Sn$, deposited on an amorphous silicon cell provides a light-trapping surface which enhances the efficiency of the device (12).

Deposition has been carried out on architectural glass yielding single-junction amorphous silicon with an efficiency of 13% in the laboratory, but with lower efficiency in production devices. An atmospheric-pressure deposition system in shown in Figure 5. The detail of the gas injection device is shown in Figure 6.

Figure 5. Atmospheric Pressure CVD Production System for the Deposition of Photovoltaic Coating (Source: Watkins-Johnson, Palo Alto, CA)

Higher efficiency is needed in order to compete effectively with conventional grid power sources. This goal may be reached as considerable R&D work constantly upgrades the technology (13).

Other Photovoltaic Materials: Several other photovoltaic materials are being actively developed. A promising system uses a multijunction in which multiple layers of optoelectronic materials are tailored to respond to specific wavelengths, thus converting a greater proportion of the light spectrum to electricity. An example incorporates an upper layer of indium phosphide (InP), which captures the visible part of the light spectrum, and a second layer, made of indium gallium arsenide (InGaAs), which captures the infra-red portion (14).

Cells made of CVD epitaxial gallium arsenide and its alloys have shown the highest efficiency (over 22% recorded against a theoretical value of 39%). Cost, however, is very high. Other materials are being investigated such as copper indium selenide (CuInSe$_2$), cadmium telluride (CdTe), indium phosphide (InP) (which has high resistance to radiation), the sulfides, selenides and tellurides of lead, selenium, germanium, and the combination zinc telluride/cadmium telluride/gallium arsenide (ZnTe,CdTe,GaAs). These materials are all deposited by MOCVD (15,16).

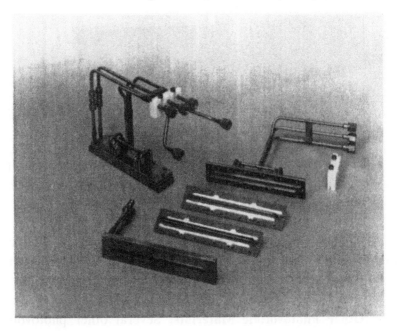

Figure 6. Detail of Gas Injector System for Atmospheric CVD of Photovoltaic Materials (Source: Watkins-Johnson, Palo Alto, CA)

5.0 CVD IN FERROELECTRICITY

Ferroelectric materials are capable of being polarized in the presence of an electric field. They may exhibit considerable anomalies in one or more of their physical properties including piezoelectric coefficient, pyroelectric coefficient, dielectric constant and optoelectronic constant. In the latter case, the transmission of light through the material is affected by the electric field which produces changes in refractive index and optical absorption coefficient. Varying the applied field changes the phase modulation (17):

Eq. (1) $\Delta n_i = \frac{1}{2} n_i^3 r_{ij} E_j$

  Δn_i = change in the refractive index

  r_{ij} = electro-optic coefficient

  E_j = applied electrical field

Important ferroelectric materials are those with piezo-electric characteristics. They are crystalline ceramics that exhibit expansion along one crystal axis and contraction along another when subjected to an electrical field. Conversely, compression generates an electrical voltage across the material. These materials have a large number of industrial applications.

Bulk ferroelectric materials are produced from powder by standard ceramic processing techniques. Until recently, thin-film ferroelectrics were deposited mostly by RF reactive sputtering, reactive ion plating, or by reactive MBE. CVD, particularly MOCVD, is now being used increasingly either experimentally or in production as the technology improves and new precursor materials become more readily available.

CVD-derived powders may prove very useful and profitable in the production of bulk ferroelectric materials which are produced by hot pressing or sintering. These powders offer great uniformity, small particle size and high reactivity.

5.1 CVD Ferroelectric Materials and Their Properties

Ferroelectric materials produced by CVD include the following:

- Lead titanate, $PbTiO_3$, has excellent pyrolectric and piezoelectric properties. Its CVD reaction is described in Chapter Seven, Section 5.11) (18).

- Lead zirconate titanate, $Pb(ZrTi)O_3$ (PZT), is a widely used compound with excellent piezoelectric properties and high Curie temperature.

- Strontium titanate, $SrTiO_3$, has a high dielectric constant (310 for single crystal) and high refractive index. It is deposited by the reaction of titanium isopropoxide with a strontium diketonate complex (19).

- Bismuth titanate, $Bi_4Ti_3O_{12}$, has a high dielectric constant and a high dielectric strength. It is deposited by the reaction of titanium isopropoxide with triphenyl bismuth (20).

- Lithium niobate, $LiNbO_3$, has piezoelectric properties (21).

- Cadmium sulfide is used as a piezoelectric film. Its CVD is similar to that of zinc sulfide described in Chapter Seven, Section 7.4.

- Aluminum nitride has a large piezoelectric coupling factor and a high surface acoustic wave velocity (5650 m/sec). Its CVD is described in Chapter Seven, Section 4.1.

- Zinc oxide is used as a piezoelectric film, with high surface acoustic wave velocity. Its CVD is described in Chapter Seven, Section 5.9.

5.2 Applications of Ferroelectric CVD Materials

The major piezoelectric applications are sensors (pick-ups, keyboards, microphones, etc.), electromechanical transducers (actuators, vibrators, etc.), signal devices, surface acoustic wave devices (resonators, traps, filters, etc.). Typical materials are: ZnO, AlN, $PbTiO_3$, $LiTaO_3$ and $Pb(Zr.Ti)O_3$ (PZT).

The major pyroelectric applications are those related to infra-red sensing such as cooking controls, fire or heat alarms, door openers, etc.. Typical materials are: $LiTaO_3$, $PbTiO_3$ and $Pb(Zr.Ti)O_3$.

REFERENCES

1 *Photonic Spectra*, 52 (Nov. 1990)

2. Heartling, G., Transparent Electrooptic PLZT Ceramics- a Process Review, *SAMPE J.*, 9-15, (Nov./Dec.1987)

3. Tuttle, B.A., Electronic Ceramic Thin Films: Trends in Research and Development, *MRS Bull.*, 40-45 (Oct./Nov. 1987)

4. Opto-electronic Devices, *Catalog 86-1 Issue III*, Lumex, Palatine, IL 60067 (1986)

5. *Comline Electronic*, Tokyo, Japan, p.2 (Jan. 1989)

6. *Japanese New Materials: Electronics*, Tokyo, Japan (Jan. 1989)

7. Woolfson, R.G., Application Specific Electronic Materials by Ion Implantatiion and CVD, *High Performance Inorganic Thin Film Coatings Conf.*, GAMI, Gorham, ME 04038 (Oct. 1968)

8. Zweibel, K., Photovoltaic Cells, *C&EN*, 34-48 (Jul.7, 1986)

9. Delahoy, A., Doele, B., Ellis, F., Ramaprasad, K., Tonon, T. and Van Dine, J., Amorphous Silicon Films and Solar Cells Prepared by Mercury-Sensitized Photo-CVD of Silane and Disilane, *Materials Issues in Applications of Amorphous Silicon Technology*, *(*D.Adler et al, Eds*)*, MRS Proc. 49:33-39 (1985)

10. Delahoy, A.E., Recent Developments in Amorphous Silicon Photovoltaic Research and Manufacturing at Chronar Corporation, *Solar Cells*, 27:39-57 (1989) ·

11. Ichikawa, Y., Sakai, H. and Uchida, Y., Plasma CVD of Amorphous Si Alloys for Photovoltaics, *Proc. 10th. Int. Conf. on CVD* (G. Cullen, Ed.) 967-976, Electrochem. Soc., Pennington, NJ 08534 (1987)

12. Gordon, R.G., Proscia, J., Ellis, F. and Delahoy, A., Texture Tin Oxide Films Produced by Atmospheric Pressure Chemical Vapor Deposition from Tetramethyltin and Their Usefulness in Producing Light Trapping in Thin Film Amorphous Silicon Solar Cells, *Solar Energy Materials*, 18:263-281 (1989)

13. Macneil, J., Delahoy, A., Kampas, F. Eser, E., Varvar, A. and Ellis, F., A 10 MW a-Si:H Module Processing Line, *Proc. 21st. IEEE Photovoltaic Specialists Conf.*, Kissimee, FL (May 21-25, 1990)

14. *Photonic Spectra*, p. 8 (Nov. 1990)

15. *Ceramic Bulletin*, 68-8:1404 (1989)

16. Chu, S.S., Chu, T., Chu, L., Han, K., Liu, Y. and Mantravadi, M., Chemical Vapor Deposition of Cadmium Telluride Films for Photovoltaic Devices, *Proc. 10th. Int. Conf. on CVD* (G. Cullen, Ed.) 982-989, Electrochem. Soc., Pennington, NJ 08534 (1987)

17. Glass, A.M., Materials for Photonic Switching and Information Processing, *MRS Bul.*, 16-20 (Aug. 1988)

18. Yoon, S.G. and Kim, H.G., Preparation and Deposition Mechanism of Ferroelectric $PbTiO_3$ Thin Films by Chemical Vapor Deposition, *J. Electrochem. Soc.*, 135- 12:3137-3140 (1988)

19. Feil, W.A., Wessels, B., Tonge, L. and Marks, T., Chemically Vapor Deposited Strontium Titanate Thin Films and Their Properties, *Proc. 11th. Int. Conf. on CVD* (K. Spear and G. Cullens, Eds.) 148-154, Electrochem. Soc., Pennington, NJ 08534 (1990)

20. Wills, L.A. and Wessels, B., Deposition of Ferroelectric $Bi_4Ti_3O_{12}$ Thin Films, *Proc. 11th. Int. Conf. on CVD* (K. Spear and G. Cullens, Eds.) 141-147, Electrochem. Soc., Pennington, NJ 08534 (1990)

21. Wakino, K., Piezoelectric and Pyroelectric Ceramics, in *Fine Ceramics* (S. Saito, Ed.) 251-261, Elsevier, N.Y. (1988)

10

CVD IN OPTICAL APPLICATIONS

1.0 INTRODUCTION

The science and technology of optics are presently in a state of rapid development. The invention of the laser and more recently that of optical storage of information are two major factors in this rapid growth. Until the development of the laser, optical coatings were relatively unsophisticated. They were produced mostly by evaporation, usually for single-layer anti-reflection applications or for superimposed layers of high and low refractive index materials for spectrally more selective filters. The laser required far more complex systems and prompted a considerable development of the thin film technology. These films are now playing a major role in most optical applications.

Evaporation, the original coating process, is relatively low cost and uncomplicated. However, evaporative coatings generally have a pronounced columnar structure which causes high light scatter, a high wavelength shift with temperature and time, and a high stress. For these reasons, evaporation is gradually being replaced by other thin-film processes, primarily sputtering. CVD techniques such as plasma CVD and MOCVD are also strong contenders and may eventually challenge and overtake both evaporation and sputtering, especially in high-quality applications or where surfaces of three-dimensional objects are to be coated. In this chapter, the role of CVD in optics is reviewed.

2.0 OPTICAL PROPERTIES

Thin films may be used to modify the optical characteristics of a system and a brief review of these characteristics is in order.

The Electromagnetic Spectrum: The optical properties of a material are related to the electromagnetic spectrum (see Chapter Nine, Figure 1). The parts of the spectrum of interest are the following:

• The visible, that is the range of wavelength that can be detected and identified as colors by the eye, extending from 0.4 to 0.7 μm.

• The infrared (IR) of wavelength immediately above the visible ranging from 0.7 to 3 μm (for present optical applications). IR radiation is a major source of heat.

• The ultraviolet (UV) with a range of wavelength immediately below the visible (from 0.19 to 0.4 μm for optical applications)

Reflectance, Transmittance, and Absorption: The optical property of reflectance is the ratio of reflected flux of light to the incident flux. Unless otherwise specified, the total reflectance is meant, which includes specular and diffuse reflectance. Transmittance is the ratio of the radiant flux transmitted by a material to the incident flux. Absorption, the attenuation of a beam through a transparent medium, can be characterized by an absorption coefficient. For thin films such as those of optical oxides, absorption is very small and can usually be ignored. In metallic or semiconductor coatings however, absorption is a major factor.

Index of Refraction: Another important characteristic of an optical material is the index of refraction which is the ratio of the velocity of light in air to the velocity of light in the material at a given wavelength. When light (including IR and UV radiations) travels in a transparent medium, such as air, and meets another medium, such as glass, a division of the light energy occurs, part being transmitted through the glass and part reflected. These optical effects are illustrated in Figure 1.

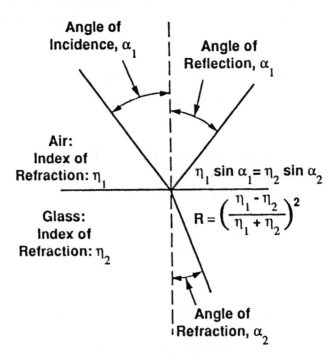

Figure 1. Schematic Diagram of Light Reflection and Refraction

3.0 OPTICAL MATERIALS PRODUCED BY CVD

Most inorganic optical coatings are of oxides, such as TiO_2, SiO_2 and ZrO_2, or nitrides such as Si_3N_4 and BN. Very thin films of metals such as aluminum, silver or gold are also used for reflecting unwanted heat or where electrical conductivity is needed.

The following is a partial list of optical materials that are presently obtained by CVD either in production or experimentally:

- SnO_2 for emissivity control (E glass)

- Very thin multilayers of oxides with high refractive index (TiO_2, ZrO_2, HfO_2, ThO_2) and low refractive index (SiO_2) for antireflection coatings

- SiO for filters (2.7 μm band)

- SnO_2 and indium tin oxide (ITO) for conductive transparent coatings on glass for electro-magnetic interference (EMI) applications

- Fe_2O_3 for beam splitters and interference layers

- Ta_2O_3 as antireflective coatings and hot mirrors

- ZnS and ZnSe for IR windows

- Amorphous silicon for photothermal conversion

- Nitrides of Ti, Zr and Hf for optical selectivity (1)

- Molybdenum thin films for IR reflectance (2)

- Tungsten thin films from $W(CO)_6$ for reflectivity control (3)

- Nitrides by plasma CVD as antireflection coatings for photovoltaic cells (4)

- Alternating layers of Si_3N_4 and SiO_2 for color filters (5)

- Diamond-like-carbon (DLC) for abrasion protection of IR optical windows

- DLC for improving resistance of thin-film mirrors to X-rays

The CVD reactions for these materials are described in Chapters Five, Six and Seven.

4.0 OPTICAL APPLICATIONS OF CVD

4.1 Antireflection Coatings

The function of an antireflection (AR) coating is to reduce the surface reflection of optical elements and increase the amount of light transmitted. The glare and ghost images from secondary reflections are minimized and the result is a cleaner image. Multiple coatings are now produced that can reduce the reflection to 0.3% or less over a broad band of frequencies. A typical AR material is magnesium fluoride, MgF_2.

Most AR coatings are still produced by evaporation but recently CVD has made large inroads particularly in applications with three-dimensional surfaces or deep recesses. AR coatings are used in numerous applications which include lenses for cameras and binoculars, instrument panels, microscopes, telescopes, range finders etc. as well as on automotive and architectural glasses.

4.2 Reflective Coatings

The purpose of a reflective coating is to provide maximum reflectivity. The major application is found in mirrors. In addition to reflection, other properties such as abrasion resistance and good adhesion to glass are often required. A common reflective material is aluminum which has excellent reflectivity. However it has poor abrasion resistance and adhesion. Because of that, it is being replaced in many critical applications by chromium, rhodium, molybdenum or tungsten which have much better abrasion resistance and improved adhesion, although their reflectivity is not as high. These metals are deposited by CVD (see Chapter Five).

Reflective coatings are found in astronomical instruments, telescopes, spectrometers, range finders, projectors, microfilm readers, television tubes, diffraction gratings, interferometers, beam splitters, and many other applications.

Reflective and antireflective coatings are also used extensively in lasers. A laser is a coherent light beam emitted over an extremely small frequency range which can be directed with great accuracy. Laser applications are very extensive and include such diverse areas as communication, information storage and retrieval, meteorology, medicine, and biological sciences.

4.3 Heat and Light Separation Coatings

Heat- and light-separation coatings (also known as hot and cold mirrors) are very important applications that operate by separating the hot (IR) from the cold (visible) radiation. The principle is shown schematically in Figure 2. Figure 3a shows the transmittance of a cold mirror that is coated with a dielectric film reflecting more than 90% of the visible spectrum while transmitting more than 80% of the IR (below 8000 μms). The hot mirror uses the opposite principle by reflecting the IR and transmitting the visible (Figure 3b). Tin oxide and indium tin oxide (ITO) are excellent IR reflectors and are both deposited by CVD (see Chapter Seven, Section 5.7).

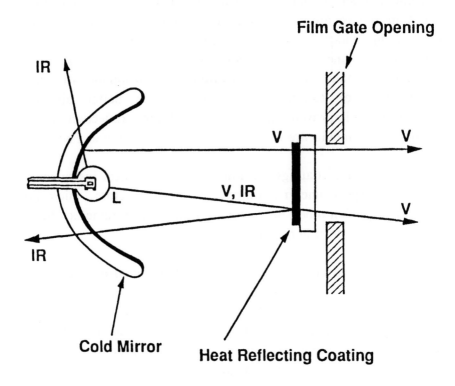

Figure 2. Schematic Diagram of Operation of Hot and Cold Mirrors

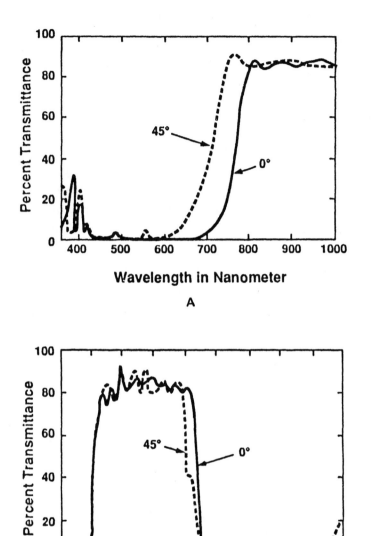

Figure 3. Reflection and Transmittance of Hot and Cold Mirrors

Hot and cold mirror CVD coatings are used in projectors to maintain the film gate at low temperature and avoid damaging the film. They are also used increasingly in tungsten-halogen lamps, a potentially very important application now on a production basis (6).

4.4 Electrically Conductive Transparent Coatings

Some metal oxides have the unique property of being electrically conductive, in contrast with the majority of oxides which are good dielectrics at least at room temperature. The most widely used conductive oxides are tin oxide (SnO_2) and indium oxide (In_2O_3). The latter is usually doped with tin oxide (90% indium, 10% tin) and is known as ITO (indium tin oxide). In the form of thin films on glass, these oxides have very reproducible electrical conductivity and high light transmission in the visible spectrum. In addition, ITO films are stable and have good resistance to abrasion. They can be acid etched to specific line geometry by standard photoresist techniques. Their properties are shown in Table 1.

These coatings are used to prevent the transmission of radio frequency interference from sources such as television or computer displays. They are also used as antistatic coatings and to absorb or reflect microwaves.

A potentially important application of electrically conductive films is in de-icing automobile and aircraft windshields. These films are presently deposited by sputtering or, in some cases, by sol-gel coating but their deposition by CVD is being developed. The coating comprises a low-resistance thin metallic layer sandwiched between two antireflection oxide layers and especially designed bus bars to distribute the electrical energy evenly. To provide erosion resistance, the coating is sandwiched within the safety glass. A 3 mm (1/8 in.) layer of ice can be removed in 2 min.

at -13°C. Also under development are reflective and low-emissivity CVD coatings for all automobile glass areas to reduce the heating and cooling load.

TABLE 1

Properties of Transparent Electrically Conductive Coatings

	Noble Metals	Tin Oxide	ITO
Thickness (Å)	350-70	3000-100	3000-100
Surface resistivity, r (ohm per square)	1-30	80-50000	7.5-1000
r-Tolerance	+/- 20%	+/- 50%	+/- 50%
Luminous transmission %	70-85	70-80	75-90
Abrasion resistance	Good	Excellent	Very good
Chemical resistance	Excellent	Excellent	Very good
Flatness	Conforms to substrate	Distorts substrate	Conforms to substrate
Adhesion	Excellent	Excellent	Excellent
Resistivity stability	Excellent	Very good	Good
Cost	Moderate	Low	Moderate

The production technology is presently dominated by sputtering, which produces an optically satisfactory product but requires it to be sandwiched within the safety glass because of poor scratch resistance. CVD coatings are is superior in this respect and may gain a foothold for that reason.

Present coatings lack abrasion resistance and their light transmission is still inadequate. Considerable effort is spent to improve these properties. The CVD oxides films shown in Table 1 are particularly attractive since they are inherently abrasion resistant and could be used on the outer surface of the glass.

4.5 Architectural Glass Coating

A major application of thin films is the coating of glass to control the effects of solar radiation in buildings. Ideally, such films should regulate the transmission and reflection of the solar spectrum as shown in Figure 4. The solar heat energy wavelengths are limited to the near infrared (up to 2 μm) while the average heat energy from a heated room is in the far infrared. As a result, a coating system should have controllable optical properties so that, in the winter, it would have a high transparency for the visible and the near infrared part of the spectrum and a high reflectivity for the far infrared. Conversely, in the summer, it would have a high transparency for the visible, a high reflectivity for the near infrared and a high emissivity for the far infrared. The present state of the art of thin-films deposition still falls short of this goal which may have to wait for the development of suitable photochromic coating materials.

Present coatings (known as low-E coatings) allow sufficient light inside but reflect a controlled amount particularly in the infra-red spectrum thereby reducing cooling loads (Figure 5). In the US, two-thin film processes are presently used to deposit these coatings:

sputtering and CVD (Sol-gel coating is also used mainly in Germany).

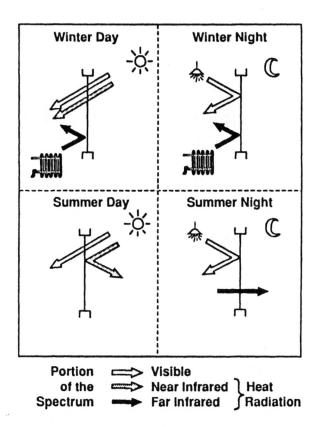

Portion   ⟹ Visible
of the   ⟹ Near Infrared ⎱ Heat
Spectrum   ➡ Far Infrared ⎰ Radiation

Winter characteristics of coating:
-Transparent to visible and near IR
- Reflective to far IR

Summer characteristics of coating:
- Transparent to visible
- Reflective to near IR
- High emissivity to far IR

Figure 4. Ideal Characteristics of Optical Coatings for Architectural Glass

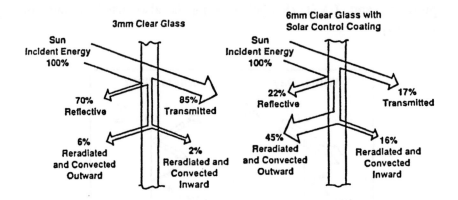

Figure 5. Performance of Present Architectural Glass Coatings

The development of large and efficient sputtering systems now allows the production of coatings with good adhesion and closely controlled thickness (to a few hundred Angstroms) which is essential to maintain a uniform level of absorption.

CVD, the other major deposition process, is used on a large scale. A typical low-E glass is obtained by depositing a very thin film of silicon dioxide followed by another very thin film of fluorine-doped tin oxide. The SiO_2 acts as a diffusion barrier and the SnO_2 reduces the emissivity. A typical CVD apparatus is shown in Figure 6 (7). These films are deposited at atmospheric pressure on the hot glass in a continuous operation (float-glass process). The CVD reactions are described in Chapter Seven, Sections 5.5 and 5.8.

4.6 Infrared Optics

Infrared optics is a fast growing area in which CVD plays a major role, particularly in the manufacture of optical IR windows. The earth atmosphere absorbs much of the infrared radiation but possesses three important bandpasses (wavelengths where the transmission is high) at 1-3 μm, 3-5 μm and 8-17 μm. As shown in

Table 2, only three materials can transmit in all these three bandpasses: single crystal diamond, germanium and zinc selenide.

TABLE 2

Properties of Infra-Red Window Materials

Material	Useful Transmission Range (wavelengths in μm)
Diamond	1-3, 3-7.5, 8-17
Diamond-like carbon	7-6
Si	1-3, 3-5
Ge	1-3, 3-5, 8-17
BaF$_2$	1-3, 3-5, 8-12
ZnS	1-3, 3-5, 8-10
ZnSe	1-3, 3-5, 8-17

Because of high cost and size limitation, single crystal diamond can be used only in exceptional cases such as the window for the Pioneer Venus spacecraft. It is however the ideal optical material as it can operate up to 800°C and is considerably harder and far less susceptible to chemical attack than the other materials.

The development of CVD production techniques may eventually allow its use on a large scale (see Chapter Six, Section 4).

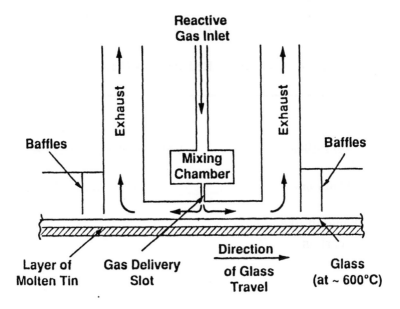

Figure 6. Schematic Diagram of CVD Apparatus for the Deposition of Solar Control Coatings on Glass

The common IR window materials are zinc sulfide which is translucent and zinc selenide which is transparent. Both of these materials are made by low-pressure CVD by the reaction of vaporized zinc and hydrogen sulfide or selenide (see Chapter Seven, Section 7). Germanium, another common IR window material, can also be made by CVD (see Chapter Six, Section 6).

These materials have the ability to transmit IR radiation with a minimum of absorption. They also have a high refractive index that does not vary to any extent with temperature or wavelength and are the chosen materials in applications where it is essential to minimize lens curvature and thickness. However, a high refractive index causes high reflectivity, as much as 36% for a germanium surface. The use of an anti-reflective (AR) coating is then necessary in order to increase the amount of light transmitted, reduce glare and ghost reflections, and obtain a clearer image. Diamond-like carbon (DLC) provides excellent AR coatings with an adjustable index of refraction which varies with the hydrogen content and can be tuned to fit a given optical design. DLC coatings are particularly well suited to germanium windows, with 90% average transmission in the 8 to 12 μm region wavelength, as shown in Figure 7, and to zinc sulfide windows, where they provide a reflectance value of 0.7% average in the 8 to 12 μm range. Adhesion to the zinc sulfide can be optimized by using an intermediate film of germanium (8). DLC coatings are deposited by CVD as described in Chapter Six, Section 4.4.

4.7 Trends in CVD Optical Applications

The following is a listing of optical applications, either present or potential, where CVD may play a major role.

• Coatings for gas-discharge light bulbs. Gradually replacing incandescent bulbs, these bulbs are far more efficient, but require a coating to reflect the IR radiation back into the filament.

• Coatings for optical storage. Most of these coatings are processed by sputtering or evaporation but MOCVD is being actively considered.

• Ultraviolet coatings for excimer lasers which are likely to play a large role in lithography of very high density computer chips.

The major optical coating process remains evaporation. However, sputtering, sol-gel coatings and CVD are making inroads because of the better properties they provide. As opposed to sputtering equipment which is very expensive and generally purchased from specialized equipment suppliers, CVD optical coaters are relatively easy to design and build and many are built in-house. At this time, it is not clear which coating process will prove best but CVD, because of the excellent bond and hardness of the deposited materials and the relatively low cost of the equipment, is a good candidate.

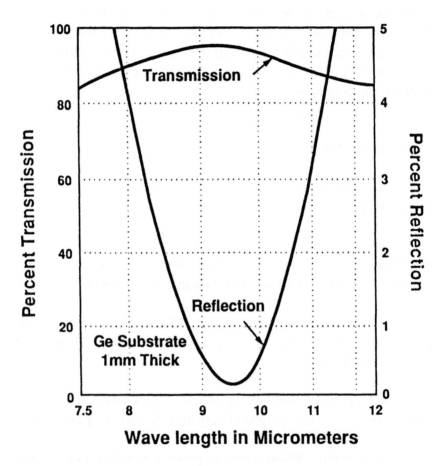

Figure 7. Transmission and Reflection of Germanium Coated with Diamond-Like Carbon (DLC)

5.0 CVD IN OPTICAL FIBER PROCESSING

5.1 Optical Considerations

Optical communications has an enormous capacity for the transmission of information and, in this respect, is far superior to communication by electrical signals. A typical system uses optical fibers as waveguides and a source of light such as a light-emitting diode (LED) or a laser in either the visible or the infrared spectrum. These fibers consist of a glass core which transmits the light, and a cladding with a lower index of refraction to keep the light within the fiber. The lower index of refraction is obtained by the addition of dopants such as boron oxide (B_2O_3) or germanium dioxide (GeO_2) during the deposition of the SiO_2. CVD is used in the production of three basic types of optical fibers. These types are shown in Figure 8 and are as follows (9):

Step-index Multimode: The step-index multimode fiber has a sharp difference between the index of refraction of the core and that of the cladding (Figure 8-A). Light propagation in such a fiber is shown in Figure 9. Only the light rays that enter the fiber within the angle of acceptance travel entirely within the core. Others, entering at a higher angle, escape the fiber. Rays entering parallel to the axis travel the shortest distance while those entering at an angle to the axis (within the angle of acceptance) travel farther and arrive later at any given point. This delay causes intermodal distortion.

Graded Multimode: In the graded-index multimode fiber, the index of refraction varies parabolically across the fiber (Figure 8-B) and is known as a "gradient index" or GRIN. In such a gradient, the speed of light (which is a function of the index of refraction) increases with the distance from the center of the core.

This greater velocity compensates for the longer path of the off-axis rays and intermoaal distortion can be avoided by careful grading of the refractive index. Needless to say, the control of the deposition parameters must be very exacting.

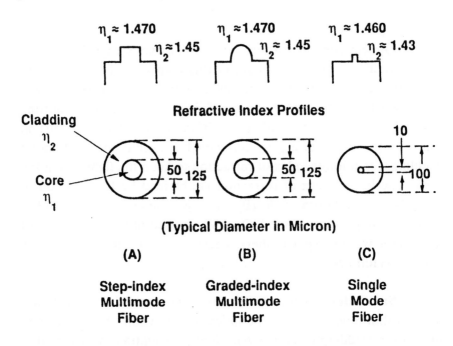

Figure 8. The Three Basic Types of Optical Fibers for Communication (9)

Single mode: The single-mode fiber has a step index profile (Figure 8-C) and a very small core diameter (typically 10 μm) such that only one mode can travel through it. This is now the preferred system, particularly for long distance. The normal light source for multimode fibers is the light-emitting diode (LED). But a LED does not have sufficient power for single mode fiber applications which, because of the smaller cross-section, require the more powerful laser. The CVD of LED and laser materials is described in Chapter Nine.

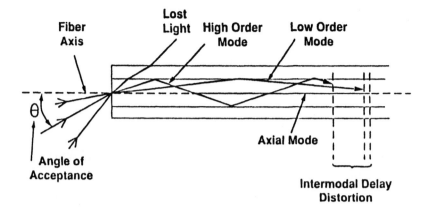

Figure 9. Light Propagation in a Multimode Step-Index Fiber

The control of impurities in the glass such as copper, iron, vanadium and hydroxyl groups, is extremely important since these are the major cause of signal attenuation. These impurities must be limited to a maximum of a few ppb which can now be achieved by CVD. Good refractive index design and impurity control are the two factors that have resulted in a considerable improvement in fiber performance. Single-mode fibers commonly transmit 400 megabit/s with a 25 km distance between repeaters and a maximum attenuation of 0.5 dB/km. Current fibers provide 250 times the information capacity of equivalent size copper wires and the standard production optical cable can now carry 1.8 million telephone calls at the same time. This is 36 times the capacity of similar optical cables produced in 1979, and is evidence of the considerable progress made in this field.

5.2 CVD Production of Optical Fibers

Two major processes are available for the production of optical fibers: direct melting (or liquid phase) and CVD (8,9). Sol-gel is being evaluated but has yet to evolve into a viable production process for that application.

The direct melt process economically produces thick optical fibers (250-400 μm in diam.) which is advantageous, but their relatively high attenuation (3-20 dB/km) due to impurities is not. As a result, they are limited to short distance multimode applications.

CVD yields fibers with very low loss (< 0.5 dB/km) which are suitable for both mono- and multimode long distance applications. It has emerged as the strongest technology for high-volume, low-cost production. Two variations of the general process are used.

The first is a classical thermal CVD method. The reactants are usually the halides, i.e. $SiCl_4$, $GeCl_4$ and $POCl_3$ with Cl_2, O_2 and $C_2Cl_2F_2$. These reactants are maintained at extremely low levels of impurities. The basic reaction to deposit the core is:

$$SiCl_4 + O_2 \longrightarrow SiO_2 + 2Cl_2$$

The same reaction is used with the dopants to form the cladding. A schematic diagram of the deposition system is shown in Figure 10. The reaction occurs inside a rotating silica tube which is heated by a traversing heat source, usually an oxy-hydrogen burner as shown in Figure 11. The SiO_2 is produced in the form of soot which deposits just ahead of the burner in a porous mass which is fused to a sintered glassy layer by the heat of the traveling burner. The cladding material is first deposited, followed by the core. The tube is then collapsed at 1800°C to form a solid preform rod which is next heated (usually in a O_2/H_2 flame) and drawn to a diameter of approximately 100 μm. The original indices of refraction of the preform are retained in the drawn fiber. The technique is used to produce step-index or graded-index multimode or single mode fibers.

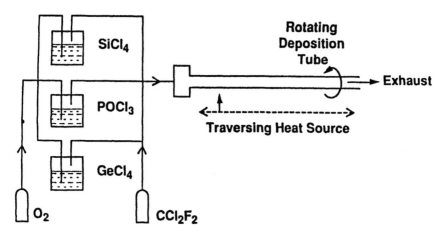

Figure 10. Schematic Diagram of Deposition Apparatus for Optical Fibers

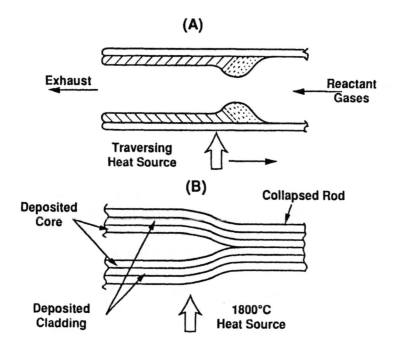

Figure 11. Deposition of Cladding (a) and Tube Collapse (b) in Optical Fiber Production

A similar deposition system uses a plasma which is produced by a traveling microwave cavity. No other source of heat is required. The deposition system is shown schematically in Figure 12. The reactants are the same as in the thermal CVD process. Pressure is maintained at approximately 1 Torr. In this case, the deposition occurs at lower temperature, no soot is formed and a compact glass is produced directly. A main advantage of this method is the more accurate grading of the refractive index of the cladding material.

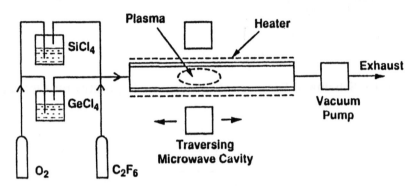

Figure 12. Schematic Diagram of Plasma CVD Apparatus for Optical Fiber Production

5.3 Infrared (IR) Transmission

Another application of fiber optics with great potential is transmission in the IR region (as opposed to the transmission in the optical region) which is being investigated extensively in the US and abroad. Glass is essentially a disordered material and as such scatters a certain amount of light due to random changes in composition and density. This scattering varies inversely with the fourth power of the wavelength. Thus, it would be advantageous to use longer wavelengths such as in the infrared. In addition, the absorption of light caused by electronic excitation is less at lower frequency (longer wavelength). Transmission in the IR region is therefore very desirable and it has been calculated that, if a suitable glass is developed, an IR signal could cross the Atlantic Ocean without amplification.

The main problem in developing fibers for IR transmission is that silica glass is not transparent in that area of the spectrum. Suitable materials include the selenides and other chalcogenide glasses (particularly for the CO_2 laser light source) and the fluorohafnate glasses. In addition, materials such as zirconium fluoride (ZrF_2), arsenic triselenide (As_2Se_3) and potassium iodide (KI) contain atoms of relatively high mass that are weakly bonded and thus transparent to IR from 2 to 5 μm. Fluoride fibers are particularly attractive because they are very flexible and ideal for medical applications such as laser angioplasty. However the CVD technique of making fibers from these materials is still in the early stage of development.

REFERENCES

1. Karlsson, B., Shimshock, R., Seraphin, B. and Haygarth, J., Optical Properties of CVD-Coated TiN, ZrN and HfN, *Sol. Energy Mater.*, 7(7):701-711 (Jan. 1983)

2. Carver, G.E. and Seraphin, B., Chemical Vapor Deposition Molybdenum Thin Films for High-Power Laser Mirrors, in *Laser Induced Damage in Optical Materials*, Publ. of National Bureau of Standards (Oct. 1979)

3. Yous, B., Robin, S., Robin, J., and Donnadieu, A., Effects of Structure on the Optical Properties of Thin Films of Tungsten Compounds, *Thin Solid Films*, 130(3- 7):181-197 (Aug. 1985)

4. Sexton, F.W., Plasma Nitride Antireflection (AR) Coatings for Silicon Solar Cells, *Solar Energy Mater.*, 7(1):1-17 (July 1982)

5. Verheijen, J., Bongaerts, P. and Verspui, G., Low Pressure Chemical Vapour Deposition of Temperature Resistance Colour Filters, *Proc. 10th. Int. Conf. on CVD* (G. Cullen, Ed.) 977-981, Electrochem. Soc., Pennington, NJ 08534 (1987)

6. IR Reflective Coatings Boosts Bulb's Output, *Photonics Spectra*, 40-41 (Jan. 1991)

7. Ellis, F.B. Jr. and Houghton, J., Chemical Vapor Deposition of Silicon Dioxide Barrier Layers for Conductivity Enhancement of Tin Oxide Films, *J. Mater. Res.*, 7-7: 863-872 (July/Aug. 1989)

8. Mirtich, M.J., The Use of Intermediate Layers to Improve the Adherence of DLC Films on ZnS and ZnSe, *J. Vac. Sci. Techno.*, 737(6):2680-2681 (Nov./Dec 1986)

9. Ketron, L.A., Fiber Optics: The Ultimate Communication Media, *Ceramic Bull.*, 66-11:1571-1578 (1987)

10. Keck, D., Low Cost Manufacture for Fibre Optic Cables, *Communications International*, 67-70, (June 1989)

11

CVD IN WEAR-, EROSION-, AND CORROSION-RESISTANT APPLICATIONS

1.0 GENERAL CONSIDERATIONS

CVD coatings are used extensively in applications requiring resistance to wear, erosion and corrosion, very often over a wide range of temperature. As mentioned in Chapter One, these coatings together with their substrates can be considered as composites which have unique combinations of properties. An example is the coating of a cutting tool such as a twist drill. The drill must be made of a tough and strong material such as high speed-tool steel that is able to withstand the stresses associated with drilling, yet its surface must be very hard and chemically resistant to withstand abrasion and corrosion. However hardness and toughness are inverse properties and no single material can have both. A solution is to coat the steel body with a refractory metal carbide or nitride which (a) protects the steel substrate from high temperature oxidation and reaction with the material to be cut and (b) provides the necessary hardness and wear resistance.

More often than not, the purpose of the coating is multifunctional. An example is the coating for an aircraft turbine blade to provide protection against the wear and erosion of a corrosive environment at high temperature as well as protection against erosion by sand and other foreign particles.

Wear-, erosion- and corrosion-resistant applications were, until relatively recently, the almost exclusive domain of the well-established techniques of hardfacing, plating and thermal spraying. These low-cost techniques are suitable for many applications which do not require the most thorough and complete protection. However, in severe environments, they are not always adequate and other surface-protection processes must be used.

Such alternative processes include boriding, nitriding, carburizing and ion implantation which modify the surface rather than coat it. The protection they afford can be more than adequate in milder environments but generally fail over a period of time if the conditions are too severe.

Another useful and widely used technique, plasma spraying, has the drawback of requiring thick deposits to insure adequate protection. Often extensive grinding and polishing are needed. Other techniques such as sputtering provide excellent protection but are limited by their line of sight characteristic, although the use of magnetron sputtering with multiple targets partially offsets this limitation (see Appendix) (1,2).

CVD suffers none of these limitations and, as a result, is being used more and more and is now an essential part of many industrial applications particularly those operating in extreme conditions. It is often the best possible solution to severe problems of erosion, friction or hot corrosion.

A special case must be made for the coating of cutting tools. The coating of cutting tools is a very important industrial application of CVD, of which much is known. It is described separately in Section 4.0.

2.0 WEAR MECHANISMS

Surface wear is defined as the deformation and loss of surface material as the result of a mechanical, thermal or chemical action. As mentioned above, these three agents can act singly but are more often found in combination which may make the wear process very difficult to analyze. Materials for wear protection have different responses to each of these wear mechanisms and, consequently, no universal wear material exists. To select the optimum material or combination of materials, it is essential to determine the cause and the mechanism of the wear as accurately as possible. The selection can then be made of the best and most cost-effective material.

2.1 Mechanical Wear

Mechanical wear is caused by the action of mechanical forces. A wear surface, viewed under high magnification, shows a very irregular topography generally characterized by a series of ridges and grooves. Contact with another surface occurs on the ridges which become highly stressed since they carry all the load. Particles on both surfaces tend to fuse together and be removed by adhesive and shear failure. Abrasive wear is a similar mechanism which results from hard particles gouging the softer surface. The wear surface may also be weakened by the applied stresses that cause fracturing along the dislocations in the lattice structure.

Beside adhesive and abrasive wear, two other wear mechanisms must be mentioned: (1) fatigue wear, which is caused by the formation and propagation of surface cracks that typically occur in ball bearings and friction gears, and (2) cavitation wear, which occurs when a solid such as ship propeller moves at high speed in a liquid, leading to the formation of bubbles which are a major cause of surface erosion.

A direct relationship is found between the hardness of a material and its ability to withstand mechanical wear: the harder the material, the greater its wear resistance. A related factor is that harder materials usually take a better polish and, as a result, the wear due to surface roughness mentioned above can be minimized.

Choosing material combinations with a low coefficient of friction is also an important factor in reducing mechanical wear.

2.2 Corrosive Wear

Corrosive wear results from a chemical reaction of the wear surface with the environment. In this section, only corrosion that occurs in conjunction with mechanical wear is considered. Purely corrosive wear is reviewed in Section 5.0 below. The chemical resistance of a given coating material must be assessed if the application involves a corrosive environment.

A typical example is the environment found in very deep oil and gas wells, (over 500 m.), which usually contain significant concentrations of CO_2, H_2S and chlorides. The corrosive effect of these chemicals is enhanced by the high temperature and pressure found at these great depths.

TiC, TiN and Si_3N_4 are usually inert to most corrosive and oxidizing environments unless the application involves high temperatures in which case oxides such as SiO_2, Al_2O_3 and TiO_2 are preferred. Yet, any coating, even if completely resistant to corrosion by itself, is of no value if pores or microcracks are present or if it is highly stressed, which makes it prone to adhesive and cohesive failure.

2.3 Temperature Effects

All materials, with a few notable exceptions such as graphite, undergo an accelerated degradation of their properties with increasing temperature. This is particularly true of chemical resistance and mechanical properties such as strength and hardness as shown for refractory carbides in Figure 1.

It should be noted that pure tungsten carbide retains its hardness up to 600°C while the hardness of most other carbides decreases rapidly with rising temperature. Diffusion of one material into the other also increases rapidly with increasing temperature, particularly in materials that have poor thermal stability.

3.0 CVD COATINGS FOR WEAR- AND CORROSION-RESISTANCE

3.1 Wear- and Corrosion-Resistance Materials and Their Properties

Both CVD and PVD are used extensively in the production of coatings for wear and corrosion resistance. PVD, especially sputtering, is the preferred process for the coating of high-speed steel tools and the deposition of nitrides such as TiN, AlN, (Ti,Al)N, CrN and BN as well as amorphous carbon (a-C:H) (see Appendix of this present book).

CVD is now used on as large a scale as sputtering. The CVD coating materials for wear and corrosion resistance are mostly the carbides and nitrides and, to a lesser degree, the borides. Table 1 summarizes certain general trends in the properties of these materials.

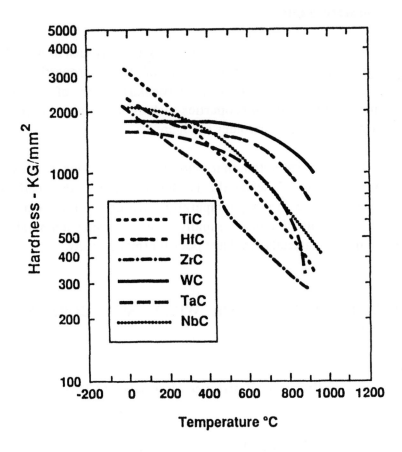

Figure 1. Hardness of Refractory Carbides as a Function of Temperature

Most coatings for resistance to erosion, wear and corrosion use a relatively small number of ceramic coating materials which are produced by CVD on an industrial scale. These materials are the following.

Titanium Carbide: TiC, with its great hardness and wear resistance, is particularly suitable to reducing mechanical and abrasive wear. However it is susceptible to chemical attack and is not a good diffusion barrier.

TABLE 1

Property Trends in Carbides, Nitrides and Borides

	High	Medium	Low
Melting Point	C	B	N
Stability	N	C	B
Thermal Expansion	N	C	B
Hardness	B	C	N
Toughness	N	C	B
Reactivity	B	C	N
Adhesion to substrate	B	C	N

C = carbides
N = nitrides
B = borides

Titanium Nitride: TiN is very stable, is an excellent diffusion barrier and has a low coefficient of friction. As such it is well suited for reducing corrosion, erosion and galling. It is used in cutting applications, as a coating on cemented tungsten carbide tools, gear components and tube and wire drawing dies.

Titanium Carbonitride: Ti(C,N) is a solid solution of TiC and TiN and combines the properties of both materials. It offers excellent protection against abrasive wear and has good lubricating characteristics. It is used to coat tools and dies for the processing of ceramics, graphite and filled plastics.

Chromium Carbide: Cr_7C_3 has excellent corrosion and oxidation resistance. It is rarely used alone but mostly in combination with TiC and TiN as a base layer .

Silicon Carbide: SiC has low thermal expansion, high hardness and is very resistant to oxidation. It is used widely in the coating of graphite and carbon to impart wear and oxidation resistance.

Titanium Boride: TiB_2 is extremely hard, corrosion resistant and provides good protection against abrasion.

Aluminum Oxide: Al_2O_3 has high strength and excellent stability and oxidation resistance. It is used extensively to coat cemented carbide cutting tools, usually in combination with TiC particularly in cutting operations which generate much heat.

Diamond-like Carbon: DLC is very hard and chemically inert but the coating is usually highly stressed in thicknesses greater than one μm. Its production applications are relatively recent but it is already used in textile machinery, bearing surfaces, measuring instruments, air bearings, precision tooling, gears, fluid engineering systems, engine components, nozzles and rotating equipment.

The characteristics and the CVD of these materials are described in Chapters Six and Seven. Their properties are summarized in Table 2.

TABLE 2

Relevant Wear and Corrosion Properties of CVD Coating Materials

	Hardness kg/mm^2	Thermal Conductivity W/cm K	Coeff. of Thermal Expansion ppm/°C at 25°C	Remarks
TiC	3200	0.17	7.6	high wear and abrasion resistance low friction
TiN	2100	0.33	9.5	high lubricity stable and inert
TiCN	2500-3000	0.2-0.3	app.0.8	stable lubricant
Cr_7C_3	2250	0.11	10	resists oxidation to 900°C
SiC	2800	1.25	3.9	high conductivity shock resistant
TiB_2	3370	0.25	6.6	high hardness and wear resistance
Al_2O_3	1910	0.34	8.3	oxidation resistant very stable
DLC	3000-5000	2.0	-	very hard, very high thermal conductivity

The comparative wear rate of some of these materials is shown in Figure 2. The wear was obtained with aluminum oxide powder in a jet abrader (3). Very hard coating materials such as TiB_2, TiC or SiC have extremely small wear rates.

The materials listed in Table 2 can be used as multilayer structures to take advantage of the strongest characteristics of each. An example is a multilayer coating for cutting tools consisting of sequential layers of TiN for lubricity and galling resistance, Al_2O_3 for chemical inertness and thermal insulation and TiC and Ti(C,N) for abrasion resistance.

Other CVD-produced materials used to a lesser extent or on an experimental basis for wear, erosion and corrosion applications include the following:

• (Ti,Al)N has essentially the properties of TiN but with much higher oxidation resistance. It is now deposited by sputtering and its deposition by CVD is being investigated.

• Boron carbide (B_4C) is extremely hard and is used where maximum resistance to erosion is required. It has good nuclear properties (see Chapter Seven, Section 3.1).

• Hafnium nitride (HfN) has good oxidation resistance and is developed as a cutting tool material (see Chapter Seven, Section 4.3) (4).

• Molybdenum disulfide (MoS_2) and tungsten diselenide (WSe_2) are used for their excellent dry lubricant properties.

• Polycrystalline diamond, the hardest material, may provide the best wear and erosion resistance material, once the deposition problems are solved and it has become commercially viable (see Chapter Six, Section 4).

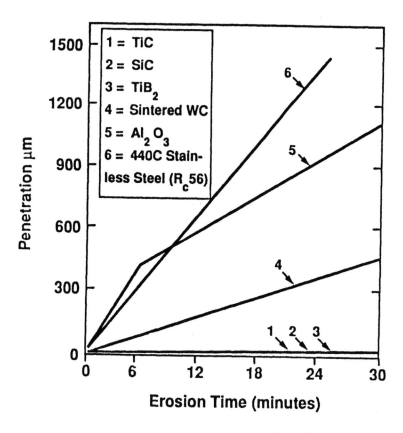

Figure 2. Comparative Wear Rate of Selected Hard Materials

3.2 Wear and Corrosion Resistance Applications of CVD Coatings

Existing production applications where CVD coatings are used for wear, erosion and corrosion protection can be classified in the broad categories listed here with some typical applications.

Metal Forming (non-cutting):

Tube and wire drawing dies (TiN)
Stamping, chamfering and coining tools (TiC)

Drawing punches and dies (TiN)
Deep drawing dies (TiC)
Sequential drawing dies (Cr_7C_3)

Ceramic and Plastic Processing:

Molding tools and dies for glass-filled plastics
[Ti(C,N)]
Extrusion dies for ceramic molding (TiC)
Kneading components for plastic mixing (TiC)

Chemical and General Processing Industries:

Pump and valve parts for corrosive liquids (SiC)
Pump and valve parts for abrasive liquids (TiB_2)
Valve liners (SiC)
Positive orifice chokes (SiC, TiB_2)
Packing sleeves, feed screws (TiC)
Thermowells (SiC, Al_2O_3)
Abrasive slurry transport (WC) (5)
Sandblasting nozzles (TiC, B_4C, TiB_2)
Textile-processing rolls and shafts (Al_2O_3, TiC, WC)
Paper processing rolls and shafts (TiC)
Valves for coal liquefaction components (TiB_2) (6)
Solder handling in printed-circuit process (TiC, TiN)

Machine Elements:

Gear components (TiN)
Stainless steel spray gun nozzles (TiC)
Components for abrasive processing (TiC)
Ball bearings (TiC) (7)
Turbine blades (SiC, TiC)

4.0 CVD IN CUTTING-TOOL APPLICATIONS

The cost of machining is estimated at over $60 billion annually in 1989 in the US alone and increasing rapidly as new materials, presenting a greater machining challenge, appear on the market. This cost forms a large proportion of the total manufacturing cost and, in some industries such as airframe manufacturing, may reach as high as 70%. This means a large market for machining equipment and for cutting tools. Almost $4 billion was spent in the US alone for cutting tools in 1989 and the forecast is for increasing expenditures.

Coatings play a vital part in the cutting-tool industry and this is where CVD technology has made some of its most important gains. As an example, CVD films of titanium carbide on cemented carbide tools were first commercialized in the early sixties and their use has continuously increased ever since. It is estimated that in the US, by 1995, over 85% of all cemented carbide tools will be coated. In 1980, the amount was only 25%. In Japan, carbide coating was adopted earlier and more readily and the percentage is even higher.

4.1 Cutting Requirements

Modern machining deals with an increasingly wide range of materials which includes, in addition to the traditional metals, high-chromium and nickel stainless steels, titanium, intermetallics, re-fractory metals, ceramics, glasses, fiber- or whisker-reinforced composites and many others. These materials have widely different properties. They react differently to machining and each presents a special machining problem.

The type of machining and its demands on the tool also vary greatly. Operations such as grinding, cutting, milling, drilling, tapping and turning, all affect the tools in different ways. With such a wide variety of materials to be machined and large differences in machining conditions, it is obvious that no single tool material is able to meet all the requirements. A tool material suitable to cut aluminum differs from one to drill a carbide. The great flexibility provided by tool coatings has been particularly useful in meeting all these requirements and has been a vital factor in the development of modern machining.

4.2 Wear and Failure Mechanisms

The wear and subsequent failure of a cutting tool is a complex mechanism that usually involves a number of physical and chemical phenomena. Temperatures at the tool/workpiece interface (cutting edge) may reach up to 1200°C in a very short period of time. This creates a pronounced thermal shock and promotes oxidation of the tool surface and the diffusion of metallic constituents of the tool into the chip with a resulting loss of tool strength.

Other important factors that contribute to failure are abrasive wear, adhesive wear, chipping and plastic deformation. Abrasive wear is the result of the abrasion caused by the chip on the rake face of the tool as it moves past the workpiece. Adhesive wear galling is caused by the adhesion of a chip to the rake face of the tool due to a high temperature chemical reaction leading to localized welding. Plastic deformation is caused by the yield of the tool material under the high pressure and temperature encountered during the cutting operation.

Other wear mechanisms are flank wear and crater wear which occur mostly with cemented carbide tools (see Section 4.4 below). Flank wear refers to the depression that is formed below the cutting edge on the side of the tool caused by the abrasive wear of the cemented carbide. TiC is particularly effective in reducing it.

Crater wear occurs in the form of small depressions on the rake face behind the point of contact of the tool with the workpiece. Diffusion of the cobalt binder into the cutting chip usually occurs with crater wear. TiN is effective in reducing both diffusion and crater wear (8).

4.3 Cutting-Tool Coating Materials

Selection of the optimum coating material (or combination of materials) depends on the type of machining operation, the material to be machined and other factors (9). Criteria for such a selection are summarized in Table 3 (10).

TABLE 3

Criteria for Selecting Coating Materials for Cutting Tools

Oxidation Resistance	1. Al_2O_3
Corrosion Resistance	2. TiN
High-Temperature Stability	3. TiC
Crater Wear Resistance	1. Al_2O_3
	2. TiN
	3. TiC
Hardness	1. TiC
Edge Retention	2. TiN
	3. Al_2O_3

TABLE 3 (cont.)

Criteria for Selecting Coating Materials for Cutting Tools

Abrasion Resistance	1. Al_2O_3
Flank Wear	2. TiC
	3. TiN
Coefficient of Friction	1. TiN
Lubricity	2. Al_2O_3
	3. TiC
Fine Grain Size	1. TiN
	2. TiC
	3. Al_2O_3

TiN is the most common coating material. It is used generally in conjunction with a very thin undercoating of TiC or Ti(CN) to promote adhesion. Al_2O_3 coatings are used in high-speed machining applications where oxidation resistance and high-temperature stability are important factors. Like TiN, it is deposited on a TiC intermediate layer. Practically all coatings are multi-layer systems combining TiC, TiN, Ti(C,N) and Al_2O_3 in various sequences.

4.4 Cutting Tool Materials and the Substrate Problem

Several basic materials are used for cutting tools.

High-Speed Tool Steel: High-speed tool steel is still the major material for cutting tools. It is inexpensive, it has high strength and toughness and can be hardened up to 70 Rc. Cutting tools have sharp edges for the purpose of shaving and generating a material curl. In order for the tool to perform properly, this edge must remain sharp. In the case of uncoated tool steel, this sharp edge is lost rapidly. For better results, the material is coated with TiN or TiC.

Tool steel has an austenizing temperature between 450 and 550°C. Heating above this transition temperature, such as would be required for coating by thermal CVD, leads to softening and requires a heat treatment sequence for re-hardening. This may cause warpage and dimensional changes, in addition to increasing cost. Thus coating by high temperature CVD is not practical or economical and essentially all coatings on tool steel are done by sputtering. Work is under way to deposit TiN at lower temperature by plasma CVD (see Chapter Seven, Section 4.6) and Ti(C,N) by MOCVD at temperature low enough to coat tool steel without degradation (see Chapter Seven, Section 3.7). The increase in the lifetime of the cutting tool as the result of coating is illustrated in Figure 3 (11).

Coated tool steel is used in tips and blanks, in milling tools, in turning and boring tools and in taps, hobs and twist drills.

Cemented-Carbide Cutting Tools: This very important group of cutting-tool materials is made from a carbide powder, which is tungsten carbide (WC) in most cases, cemented under high pressure and temperature with a binder, usually cobalt. Cobalt-bonded WC has high-temperature strength and can be coated by thermal CVD with generally no problems although too high a deposition temperature may lead to binder diffusion and the formation of a detrimental tertiary carbide Co_6W_6C, called the "eta phase", at the WC/matrix interfaces. This leads to loss of strength and adhesion failure of the coating. As mentioned previously, most cemented carbides are now coated by CVD, usually with multiple coatings.

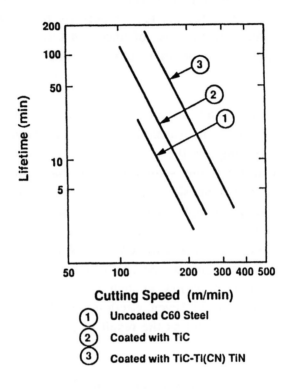

① Uncoated C60 Steel

② Coated with TiC

③ Coated with TiC-Ti(CN) TiN

Figure 3. Lifetime of Coated and Uncoated Cutting Tool Steel

The main applications of coated cemented carbides are tips and blanks, indexable inserts, milling tools, turning and boring tools and circular saws.

Ceramic Cutting Tools: The use of ceramic cutting tools is increasing due to their great deformation resistance at higher machining speed and good thermal stability. Materials include alumina, silicon nitride, Si-Al-O-N, alumina-carbide composites and, more recently, a composite of silicon nitride reinforced with silicon carbide whiskers. This last material can be produced by chemical vapor infiltration (CVI) and has very high strength and toughness as shown in Table 4 (12).

TABLE 4

Selected Properties of Commercial Ceramic Cutting Tools

Material	Hardness GPa (KHN)	Fracture Toughness MPa $\sqrt{m}$	Rupture Modulus MPa	Thermal Conduct. W-m K	Coeff. of Thermal Expansion ppm/°C
Al_2O_3*	15.6	2.9	280	32.3	8.22
Al_2O_3/TiC**	16.2	3.5	450	n/a	8.7
Al_2O_3/SiC***	17.0	6.0	675	35.2	7.35
Si_3N_4/20% SiC whiskers ****	13.0	6.0	1050	39.5	3.0

*   Kennmetal designation K060

**   NGK designation NTK-HC2

***   Greenleaf designation WG-300

****   GTE Valenite designation Quantum 6

These ceramics are usually coated by CVD with an appreciable increase in performance as shown in Figure 4. A major use of these coated ceramics is in the machining of hard materials such as cast iron and superalloys (13).

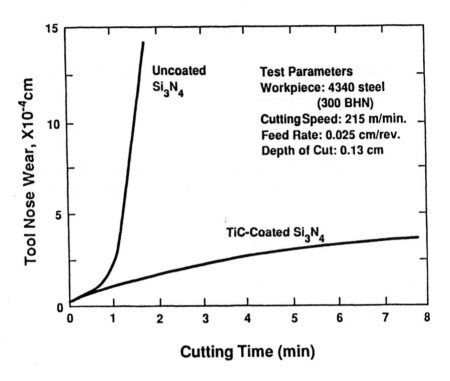

Figure 4. Wear of Coated and Uncoated Ceramic Tools

Diamond Cutting Tools: Single crystal diamond is the best all around cutting material but is very expensive and requires excellent machine conditions such as closely controlled speed and absence of chatter and vibration for optimum performance. However diamond has limited oxidation resistance and chemical stability. Diamond begins to oxidize in dry-grinding operations between 500 and 700°C and the use of coolant is recommended to remain below these temperatures.

In addition, diamond reacts at approximately 1000°C with elements which are carbide formers such as iron, cobalt, nickel, aluminum, tantalum and boron and is generally considered unsuitable to machine these materials. In such cases, a better choice is cubic boron nitride (14). However, for the cutting of very abrasive materials, diamond is usually the best choice.

Polycrystalline CVD diamond films are now being considered in cutting and grinding applications (see Chapter Six, Section 4). They are potentially very promising but such applications are still mostly in the R&D stage. One such application is the deposition of a 10 μm thick diamond film on a cobalt bonded sintered tungsten carbide base. Adhesion to the substrate is satisfactory if an intermediate carbide former layer is used. Such a film has improved the flank wear of tools by a factor of two to five (15). Initial use is for cutting aluminum and reinforced plastic. Also being tested is a diamond coated whetstone for precision grinding with a 5 to 10 μm film also using microwave deposition. Polycrystalline diamond films have extremely fine submicron crystals and, as such, are ideal for precision grinding and for very close tolerance machining.

5.0 CVD IN CORROSION APPLICATIONS

In many respects, the problems of corrosion are similar to those of wear and erosion reviewed above and often a corrosion application also requires resistance to wear and erosion. In this Section, CVD applications where corrosion resistance is the major factor will be reviewed.

Corrosion is defined as the wearing of a surface by chemical action. It is usually a slow and often unnoticed phenomenon yet it is all pervasive and incredibly costly. In the US alone, the corrosion of metals is estimated to cost over ten billion dollars a year for replacement, repairs and maintenance. To this, the cost of idled equipment and personnel must be added.

Protecting a surface from corrosion by coating can be accomplished by a number of well-established processes which include paints, the plating of metals such as zinc or cadmium, diffusion, thermal spraying and, more recently, vapor deposition.

Physical vapor deposition (PVD) is used extensively in corrosion protection, in such applications as the aluminum metallizing of steel or plastics by evaporation, the alloy coating of turbine components by electron-beam evaporation and sputtering and the coating of watches, frames for eyeglasses and similar items with titanium nitride by sputtering.

CVD is relatively a newcomer in the area of corrosion protection but is now making rapid strides and may soon become a serious competitor to PVD and other coating processes.

5.1 CVD Metals for Corrosion Resistance Applications

The range of chemical reactivity of metals is very wide, from the inertness of the platinum group to the extreme reactivity of some alkali metals. The order of metal reactivity follows essentially the order of the electrochemical series which is shown in Table 5 for the metals commonly deposited by CVD.

The metals above hydrogen react with water to form hydrogen, at room temperature for the alkali metals, at 100°C with steam for magnesium and zinc, and at red heat for manganese and cobalt.

TABLE 5

Electrochemical Series of the CVD Metals

Metal	Electrode Potential (V)
Magnesium	-2.38
Aluminum	-1.67
Titanium	-1.63
Manganese	-1.185
Niobium	-1.09
Zinc	-0.763
Tantalum	-0.750
Chromium	-0.557
Iron	-0.44
Cadmium	-0.402
Cobalt	-0.28
Nickel	-0.23
Molybdenum	-0.20
Tin	-0.141
Tungsten	-0.119
(Hydrogen)	(0)
Rhenium	0.300
Copper	0.343
Silver	0.799
Palladium	0.82
Platinum	1.118
Iridium	1.156
Gold	1.42

Oxidation is the largest cause of corrosion of metals. The conversion of a metal into an oxide by the direct combination with oxygen is the most common form of oxidation. The best known example is the rusting of steel. Oxidation, in the more general context, is the result of the increase of positive charge of an ion. Reduction is the reverse of oxidation. The most common oxidizing agents are oxygen, ozone, peroxides, the oxy-acids such as nitric, chromic and chloric acids, the halogens and many others.

The resistance of metals to oxidation varies considerably. Gold and most metals of the platinum group do not react at room temperature or at least the reaction rate is extremely slow. Copper, lead, tin, iron, chromium and aluminum are oxidized slowly and a continuous very thin film of impermeable oxide is formed which essentially prevents further oxidation. The oxide formed on iron, zinc and magnesium is discontinuous and oxidation continues linearly.

The rate of oxidation increases with temperature. Oxygen diffusion becomes more rapid, the oxide film becomes thicker and eventually cracks and spalls off due to internal stress. Oxidation then becomes continuous and rapid.

5.2 CVD Borides for Corrosion-Resistance Applications

Borides are relatively inert, especially to non-oxidizing reagents. They react violently with fluorine, often with incandescence. Reaction with other halogen is not as violent and may require some heat. Resistance to oxidation, acids and alkalis is summarized in Table 6. In oxidation conditions, a layer of boric oxide is formed on the surface which passivates it to some degree. Boric oxide melts at 450°C and vaporizes at 1860°C. It offers good protection up to 1500°C in a static environments but it has low viscosity at these temperatures and tends to flow under stress and the protection it offers is limited (16,17).

TABLE 6

Selected Chemical Properties of Transition-Metal Borides

Boride	Approx. Oxidation Threshold (°C)	Resistance to Acids			Resistance to Alkalis
		HCl	H_2SO_4	HNO_3	
HfB_2	750	poor	poor	poor	poor
Mo_2B_5	500	poor	good	poor	poor
NbB_2	500	good	good	good	poor
TaB_2	500	good	good	good	poor
TiB_2	550	poor	poor	poor	poor
W_2B_5	500	good	good	poor	poor
ZrB_2	650	poor	poor	poor	poor

Borides are generally resistant to molten metals, at least to those that do not readily form borides, such as copper, zinc, magnesium. aluminum, tin, lead and bismuth. TiB_2 is especially resistant to molten aluminum and, as such, is used in crucibles for evaporation of the metal.

5.3 CVD Carbides for Corrosion-Resistance Applications

The refractory carbides have very high melting points and low vapor pressures and are very stable in vacuum or inert atmosphere. They are generally not attacked by hydrogen or nitrogen even at high temperature. They react with the halogens at high temperature. Reaction threshold temperature for B_4C and SiC in fluorine is 700°C.

TiC has good resistance to sulfuric acid (2). A passivating oxide layer is formed up to a potential of 1.8 V at which point corrosion becomes severe. TiC is also very resistant to sea water, neutral industrial waste waters and human sweat. Cr_7C_3 is even more corrosion resistant and is used widely as a passivation interlayer.

Carbides oxidize readily although less rapidly than the nitrides but more so than the borides. Oxidation becomes more rapid going from the Group IV carbides (TiC, ZrC, HfC) to those of Group VI (Cr_3C_7, MoC, WC). In some cases, a protective film of the metal oxide is formed. Such is the case with SiC, as reviewed in Section 5.7 below.

The carbides are generally not resistant to molten slags and fused salts. Their resistance to molten metal is usually poor. For instance, TiC is attacked by nickel, cobalt, chromium and silicon. SiC is attacked by aluminum.

5.4 CVD Nitrides for Corrosion-Resistance Applications

The thermal and chemical stability of the refractory metal nitrides is high with those of Group IV (titanium, zirconium, hafnium). It decreases with those of Group V (vanadium, niobium and tantalum) and is low with those of Group VI (chromium, molybdenum and tungsten). CVD TiN, ZrN and HfN are the most commonly used. They provide a good barrier to diffusion. Their resistance to acids is generally good, at least at room temperature. However, they are decomposed by boiling alkali solutions or fused alkali. Their oxidation resistance is poor. TiN is very resistant to sea water and human sweat. It has a pleasing gold appearance and for that reason is used extensively in jewelry and similar applications.

Other useful refractory nitrides for corrosion protection are silicon nitride (Si_3N_4) and boron nitride (BN). Silicon nitride has good corrosion resistance and is not attacked by most molten metals as shown in Table 7 (see Chapter Seven, Section 4.5) (18).

TABLE 7

Resistance of Silicon Nitride to Molten Metals

Metal	Temperature (°C)	Time (hr)	Degree of Attack
Al	800	950	none
Al	1000	100	none
Pb	400	144	none
Sn	300	144	none
Zn	550	500	none
Mg	750	20	slight
Cu	1160	7	severe

Boron nitride is one of the most outstanding corrosion resistant materials. It is inert to gasoline, benzene, alcohol, acetone, chlorinated hydrocarbons and other organic solvents. It is not wetted by molten aluminum, copper, cadmium, iron, antimony, bismuth, silicon, germanium nor by many molten salts and glasses. It is used extensively as crucible material, particularly for molten metals, glasses and ceramic processing (see Chapter 7, Section 4.2).

5.5 CVD Oxides for Corrosion-Resistance Applications

The main characteristic of refractory oxides is their excellent resistance to oxidation environments. Their brittleness however makes them prone to thermal shock with the exception of silica which has a compensating low coefficient of thermal expansion. The chemical resistance of the major oxides deposited by CVD is rated in Table 8 (18).

TABLE 8

Chemical Properties of CVD Refractory Oxides

			Stability to		
	Reducing Atm.	Carbon	Acid Slags	Basic Slags	Molten Metals
Alumina	good	fair	good	good	good
Hafnia	good	fair	good	poor	good
Silica	poor	poor	good	poor	poor
Titania	poor	-	good	poor	-
Zirconia	good	fair	good	poor	good

5.6 CVD Silicides for Corrosion-Resistance Applications

The major characteristics of some silicides, particularly $MoSi_2$ and to a lesser degree $TiSi_2$ and WSi_2, is their excellent oxidation resistance. The formation of a strong adherent oxide layer protects the silicide from further oxidation, up to 1900°C in the case of $MoSi_2$ (see Chapter Seven, Section 6.1). This characteristic has been used to good advantage in the design and manufacture of heating elements for high-temperature furnaces. Element life of several hundred hours at 1800-1900°C is reported.

$MoSi_2$ is stable in SO_2, CO_2, N_2O and hydrocarbons up to 1000°C and in N_2 and CO_2 up to 1500°C. Resistance to acids is generally low.

5.7 Examples of Corrosion Protection by CVD

Oxidation Protection of Carbon: Aerospace materials are now being operated at or near their capacities with regard to temperature, stress and environment, especially in applications such as turbine-engine components. Their operational limits must be extended in order for performance improvements to be realized. The material most likely to be able to meet the new performance requirements is carbon either hot pressed or especially in the form of carbon-carbon composites. Carbon-carbon retains its strength at high temperature, has very small thermal expansion and is able to withstand the extreme thermal shock caused by high heating rates. Carbon however oxidizes rapidly above 600°C and an oxidation protection system must be used.

State-of-the-art oxidation protection systems are based on silicon carbide (SiC), which is normally applied by pack cementation or by chemical vapor infiltration (CVI) (see Chapter Four, Sections 2.0 and 9.0 and Chapter Seven, Section 3.5). Boron, zirconium, and other materials are sometimes added to the matrix of the composite or as interlayers to impart self-healing characteristics and extend the maximum operating temperature. During oxidation, a vitreous silicon dioxide (SiO_2, silica) layer is formed which provides good oxidation protection, but only to approximately 1500°C.

Any operation above that temperature is likely to result in failure, even after a short period of time. This is due to the decomposition of the SiC which may occur well below its melting point and to the formation of the volatile suboxide SiO, which evolves rapidly above 1600°C at low oxygen partial pressures by simple gasification. Another reason is the rapid decrease in viscosity of the SiO_2 protective layer with increasing temperature. This would be a severe problem with components subjected to high dynamic loads, such as a high-velocity gas stream in the case of turbine blades. A conservative design for extended life dictates an upper temperature limit of about 1370°C for a silicon carbide-based oxidation protection system.

A refractory coating has been developed that is capable of providing oxidation protection for carbon-carbon composites to 1800°C for extended periods (hours) and to 1900°C for shorter periods, thus substantially raising the effective upper temperature limit for carbon-carbon. This coating is a multilayered structure of mixed hafnium carbide and silicon carbide obtained by CVD, formed by alternating the deposition of HfC and SiC at regular intervals (see Chapter Seven, Section 3). During oxidation, a thin adherent layer of a mixed oxide is formed on the coating which provides an effective barrier to oxidation of the substrate. This mixed oxide is composed of hafnium oxide (HfO_2) and silicon oxide with hafnium silicide (16).

Other refractory oxides that can be deposited by CVD have excellent thermal stability and oxidation resistance. Some like alumina and yttria, are also good barriers to oxygen diffusion providing that they are free of pores and cracks. Many however are not, such as zirconia, hafnia, thoria and ceria. These oxides have a fluorite structure, which is a simple open cubic structure and is particularly susceptible to oxygen diffusion through ionic conductivity. The diffusion rate of oxygen in these materials can be considerable (19).

Iridium Coating for Spacecraft Rocket Nozzles: The coating of rocket nozzle with iridium is a good example of the ability of CVD to provide a complete composite material, in this case a structural refractory shell substrate coated with a corrosion- and oxidation-resistant component. The device is a thruster rocket nozzle for a satellite. The rocket uses a liquid propellant which is a mixture of nitrogen tetroxide and monomethyl hydrazine. The nozzle of original design was fabricated from a niobium alloy coated with niobium silicide and could not operate above 1320°C. This was replaced by a thin shell of rhenium protected on the inside by a very thin layer of iridium. The iridium was deposited first on a disposable mandrel, from iridium acetylacetonate (pentadionate) (see Chapter Five, Section 8.0). The rhenium was then deposited over the iridium by hydrogen reduction of the chloride (see Chapter Five, Section 14.0). The mandrel was then chemically removed. Iridium has a high melting point (2410°C) and provides good corrosion protection for the rhenium. The nozzle was tested at 2000°C and survived 400 cycles in a high oxidizer to fuel ratio with no measurable corrosion (20).

Other CVD Corrosion-Resistance Applications: The following is a listing of typical applications where CVD is used either in production or experimentally for corrosion protection.

• CVD SiO_2 provides effective protection for stainless steel in CO_2 up to 1000°C (22).

• SiO_2 and Si_3N_4 are deposited on steel by plasma CVD at low temperature on a continuous basis for oxidation protection in combination with ion plating and sputtering (21).

• Corrosion and sweat resistant CVD coatings are applied on jewelry, eye glasses and similar items in pleasing colors such as gold (TiN), metallic shine (CrN, TaN and WN) and black anthracite (C).

• Molybdenum heat pipes, coated with CVD SiC, can operate in the temperature range of 830-1130°C in an oxidizing atmosphere (23).

• Copper plates coated with TiB_2 by CVD have shown good resistance to corrosion by sea water and erosion by sand (24).

6.0 CVD IN NUCLEAR APPLICATIONS

6.1 Nuclear Fission Applications

The protection of components against nuclear radiation is a very critical factor in the design of fission nuclear systems. It is often regarded as the Achilles' heel of nuclear power. CVD is used extensively in this area, particularly in the coating of nuclear fuel particles such as fissile U-235, U-233 and fertile Th-232 with pyrolytic carbon. The carbon is deposited in a fluidized-bed reactor (see Chapter Six, Section 3.3 and Chapter Four, Section 10.0). The coated particles are then processed into fuel rods which are assembled to form the fuel elements.

The function of the carbon coating is to contain the by-products of the fission reaction, thereby reducing the shielding requirements. It also protects the nuclear fuel from embrittlement and corrosive attack and from hydrolysis during subsequent processing steps. CVD coatings of alumina deposited at 1000°C and beryllia deposited at 1400°C have also been studied for that purpose (25).

6.2 Nuclear Fusion Applications

CVD is used in many experimental coatings for fusion devices. Refractory materials with very high chemical stability and low atomic number are preferred, such as TiB_2, TiC, SiC, carbon and boron. These materials must be able to withstand very severe thermal shock (26, 27). The following applications have been reported in the literature.

- TiC coating on graphite for limiters and neutral beam armor

- Boron and B_4C deposited by plasma CVD on graphite for wall armor protection in fusion reactors(28)

- *In situ* deposition of carbon (i.e. within the fusion reactor chamber which becomes the CVD reactor) (27)

- Coatings for inertial confinement microsphere targets comprised of multilayers of Be, Au, Pt and Ta (29)

7.0 CVD IN BIOMEDICAL APPLICATIONS

Materials used in body implants must (1) meet several essential requirements such as tissue compatibility, enzymatic and hydrolytic stability and (2) have certain chemical and mechanical properties. They must not be toxic, or the surrounding tissue will die. They must be chemically resistant to the body fluids which usually have a high percentage of chloride ions. They must be biologically active if an interfacial bond is to be achieved. In some cases, they must be able to withstand continued high mechanical stresses for many years (30).

Biomedical materials include ceramics such as the biologically active hydroxyapatite and tricalcium phosphate, and high-strength metals such as titanium alloys (31). These materials are not produced by CVD as this time, except on an experimental basis.

CVD, however, is the major process used in the production of another very important biomedical material, i.e. isotropic carbon (32). In fact, more implants are made from isotropic carbon than from any ceramic material.

Isotropic carbon is obtained by the pyrolysis of a hydrocarbon, usually methane, at high temperature (1200-1500°C) in a fluidized bed on a graphite substrate (33). Under these conditions, a turbostratic structure is obtained which is characterized by very little ordering and an essentially random orientation of small crystallites.

In contrast to graphite which is highly anisotropic, such a structure has isotropic properties (see Chapter Six, Section 3). Isotropic carbon is completly inert biologically. Its properties are compared to alumina, another common implant material, in Table 9. Notable is its very high strain to failure (32).

The major biological application of isotropic carbon is in heart valves. The material is performing well and several hundred thousand units have been produced so far. Other applications include dental implants, ear prostheses, and a coating for in-dwelling catheters.

TABLE 9

Properties of Isotropic Carbon vs. Alumina

Properties	Isotropic Carbon	Alumina
Density, g/cm^3	2.1	3.89
Hardness, kg/mm^2	240-370	1500
Flexural strength, MPa	350	379
Young's modulus, GPa	28	372
Strain to failure, %	1.2	0.07-0.15
Fracture Toughness, MPa$\sqrt{m}$	2.5	4.2
Thermal expansion, ppm/°C	4-6.5	3.2-8.3
Thermal conductivity, W/cm.°C	0.032	0.34
Electrical resistivity ohm-cm	0.002	$> 10^{14}$

REFERENCES

1. Garg, D., Dyer, P., Dimos, D., Sunder, S., Hintermann, H. and Maillat, M., Low temperature CVD Tungsten Carbide Coatings for Wear/Erosion Resistance, *Ceram. Eng. Sci. Proc.*, 9(9-10):1215-1222 (1988)

2. Hintermann, H., Coatings and Solid Lubricant Films for Extreme Conditions, in *Engineering Ceramics*, 54-59, Sterling Publications, London (1988)

3. CVD Wear Coating Technical Data Sheet, *Materials Technology Corp.*, Dallas, TX 75247 (1988)

4. Oakes J., A Comparative Evaluation of HfN, Al_2O_3, TiC and TiN Coatings on Cemented Carbide Tools, *Thin Solid Films*, 107:159-165 (1983)

5. Bhat, D. and DeKay, Y., Comparison Between Laboratory Characterization and Field Performance of Steel Mud Pump Liners Coated with CM500L, a Tungsten Carbon Alloy, *ASTM Special Tech. Publ.* 946, 103-116, Philadelphia, PA 19103 (1987)

6. Curlee, R.M., Materials Coatings for Valves Coal Liquefaction Components and Instrumentation, *SAND82- 2137, Sandia National Laboratories*, Albuquerque, NM 87185 (1982)

7. Boving, H., Hintermann, H. and Hanni, W., Ball Bearings with CVD-TiC Coated Components, *Proc. 3rd. Europ. Space Mechanisms and Tribology Symp.*, Madrid, Spain, ESA SP-279 (Dec. 1987)

8. Schintlmeister, W., Wallgram, W. and Kanz, J., Properties,

Applications and Manufacture of Wear- resistant Hard Material Coatings for Tools, *Thin Solid Films*, 107:117-127 (1983)

9. Stjernberg, K. and Thelin, A., Wear Mechanisms of Coated Carbide Tools in Machining of Steel, *Proc. ASM Int. Conf. on High Productivity Machining, Materials and Processing*, Paper No. 8503-004, ASM, Metals Park, OH 44073 (May 1985)

10. Bhat, D. and Woerner, P., Coatings for Cutting Tools, *J. of Metals*, 68-69 (Feb. 1986)

11. Schintlmeister, W., Wallgram, W., Kanz, J. and Gigl, K., Cutting Tool Materials Coated by Chemical Vapour Deposition, *Wear*, 100:153-169 (1984)

12. Wayne, S.F. and Buljan, S.T., The Role of Thermal Shock on Wear Resistance of Selected Ceramic Cutting Tool Materials, *Ceram. Eng. Sci. Proc.*, 9(9-10):1395-1408 (1988)

13. Bhat, D., Techniques of Chemical Vapor Deposition, in *Surface Modification Technologies - An Engineer's Guide*, (T. Sudarshan, ed.), Marcel Dekker (1989)

14. Gardinier, C.F., Physical Properties of Superabrasives, *Amer. Ceram, Soc. Bul.*, 67-6:1031-1036 (1988)

15. Orr, R., The Impact of Thin Film Diamond on Advanced Engineering Systems, in *Proc. of Conf. on High Performance Inorganic Thin Film Coatings*, GAMI, Gorham, ME 04038 (1988)

16. Pierson, H.O., Sheek, J.G and Tuffias, R.H., Overcoating of Carbon-carbon Composites, *Wright Research and Development Center*, WRDC-TR-89-4045, Wright Patterson AAFB, OH 45433 (1989)

17. Kaufmanm, L. and Clougherty, E.V., Investigation of Boride Compounds for High Temperature Applications, in *Proc. 5th Plansee Seminar*, 722-738, Reutter, Austria (1965)

18. Campbell, I. and Sherwood, E., *High Temperature Materials and Technology*, John Wiley & Sons, New York (1967)

19. Smith, A.W., Meszaros, F. and Amata, C., Permeability of Zirconia, Hafnia and Thoria to Oxygen, *J. of the Amer. Ceram. Soc.*, 49-5:240-244 (1966)

20. Harding, J.T., Kazarof, J.M. and Appel, M.A., Iridium Coated Thrusters by CVD, NASA T.M. 101309, in *Proc. of Second Int. Conf. on Surface Modification Technologies*, AIME and ASM, Chicago, IL (Sept. 1988)

21. Hashimoto, M., Ito, S. and Ito, I., Combined Deposition Processes Create New Composites, *Research and Development*, 112-114 (Oct. 1989)

22. Bennett, M.J., Houlton, M.R. and Hawes, R.W., The Improvement by a CVD Silica Coating of the Oxidation Behaviour of a 20% Cr - 25% Ni, Niobium Stabilized Stainless Steel in Carbon Dioxide, *Corros. Sci.*, 22(2):111-133 (1982)

23. Merrigan, M., Dunwoody, W. and Lundberg, L., Heat Pipe Development for High-Temperature Recuperator Application, *J. Heat Recovery Syst.*, 2(2):125-135 (1982)

24. Motojima, S. and Kosaki, H., Resistivities Against Sea- Water Corrosion and Sea-Sand Abrasion of TiB_2 Coated Copper Plates, *J. Mater. Sci. Lett.*, 4(11):1350-1352 (Nov. 1985)

25. Oxley, J.H., Nuclear Fuels, in *Vapor Deposition* (C. Powell, J. Oxley and J. Blocher Jr. Eds.), 484-505, John Wiley & Sons, New

York (1966)

26. Mullendore, A.W., Whitley, J.B., Pierson, H.O. and Mattox, D.M., Mechanical Properties of Chemical Vapor Deposited Coatings for Fusion Reactor Application, *J. Vac. Sci. Technol.*, 18(3):1049-1053 (Apr. 1981)

27. Smith, M.F. and Whitley, J.B., Coatings in Magnetic Fusion Devices, *J. Vac. Sci. Technol.*, A4(6):3038-3045, (Nov.Dec. 1986)

28. Groner, P., Gimzewski, J.K. and Veprek, S, Boron and Doped Boron First Wall Coatings by Plasma CVD, *J. Nucl. Mat.*, 103(1-3):257-260 (1981)

29. Meyer, S.F., Metallic Coating of Microspheres, *J. Vac. Sci. Technol.*, 18(3):1198 (April 1981)

30. Hench, L. and Wilson, J., Surface Active Biomaterials, *Science*, 226:630-635 (1984)

31. Boretos, J.W., Advances in Bioceramics, *Advanced Ceram. Materials,* 2(1):15-22, (1987)

32. Beavan, A., LTI Pyrolytic Carbon for Superior Wear and Durability, *Materials Engineering,* 39-41 (Feb. 1990)

33. Bokros, J.C., Carbon in Medical Devices, in *Ceramics in Surgery*, (P. Vincenzini, ed.), Elsevier (1983)

CVD IN FIBER, POWDER, AND MONOLITHIC APPLICATIONS

1.0 INTRODUCTION

The applications of CVD considered in the previous four chapters are related to coatings that are used to modify the surface properties of the substrate. In this chapter, the CVD applications which are not directly related to coatings are reviewed. They can be divided into the following categories: fibers, powders, and monolithic shapes and composite structures. It is estimated that, in 1990, coatings formed over 85% of the applications of CVD and the balance was shared more or less equally by each of these three categories (1). This percentage, while small, is not negligible and may well increase considerably as some of the present experimental applications become commercially viable.

2.0 CVD IN FIBER APPLICATIONS

CVD is used on a large scale in the production of inorganic structural fibers such as boron and silicon carbide. Boron fibers are in fact one of the earlier commercial applications of CVD since the feasibility of the process and the superiority of the product were demonstrated as far back as 1959 (2). CVD is also being considered for the production of several other types of metal and ceramic fibers.

The process competes with the traditional method of fiber production in which the precursor materials are melted, usually in an arc furnace, drawn through spinnerets and spun or impinged by high pressure air. The melt-spin process is not well suited to materials with high melting points such as zirconia, silicon carbide or pure alumina.

A more recent competitor is sol-gel process where the fibers are produced from a chemical solution, usually of an alkoxide or mixed alkoxides, which is partially thickened by polymerization or by additives. The solution is spun directly into fibers or extruded through spinnerets and the resulting fibers are dried and sintered. Sol-gel does not require the high temperature of the direct melt process and is being used successfully in the commercial production of several types of mixed oxide fibers based on alumina and in the production of silicon carbide fibers (3). Sol-gel is the main competitor to CVD for the production of high-temperature structural fibers. Each process has its own advantages and drawbacks and, at this stage of development, it is not possible to forecast which one will prevail. It is likely that sol-gel will be the preferred process for oxide fibers and CVD for carbide, nitride and boride fibers.

2.1 The CVD Process for Fiber Production

The CVD process presently used to produce boron and silicon carbide fibers requires a monofilament starter core capable of being heated resistively such as a tungsten or graphite fiber. The deposition apparatus is shown schematically in Figure 1 (2). Since a core is required, it is not possible to produce fibers of very small diameter and present production diameters are in the range of 100-150 μm, compared to an average of 10-20 μm for sol-gel derived fibers. Being very stiff with a high Young's modulus and large diameter, they cannot be bent to a small radius or along compound shapes. They are difficult to weave into fabrics. Their use is limited to parts having a simple geometry such as plates, rods or tubes.

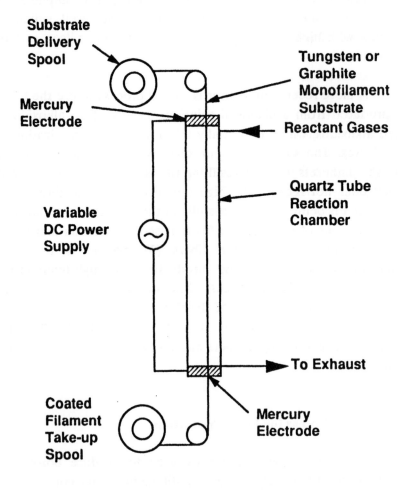

Figure 1. Boron and SiC Fiber Deposition Apparatus

A new process based on laser CVD does not require a core material and is able to produce fibers with a much smaller diameter. Deposition rate is up to 1 mm/sec. The process is still experimental and is presently being developed for the production of boron, SiC and Si_3N_4 (4). No information is given in Reference 4 as to how the core-less deposition is accomplished. It could be done by impinging the laser beam on the growing end of a retreating fiber in a CVD atmosphere.

2.2 The CVD of Boron Fibers

Processing: The CVD reaction used to produce boron fibers is the hydrogen reduction of boron chloride (see Chapter Six, Section 2.1) (5). At the deposition temperature of 1300°C, the tungsten core reacts rapidly with the deposited boron to form the brittle tungsten borides W_2B_5 and WB_4. No metallic tungsten remains. The "boron" fiber actually consists of a boron envelope surrounding a tungsten boride core which typically occupies 5% of the fiber cross section.

Properties: CVD boron fibers have high strength, high modulus and low density. Their properties are summarized and compared with SiC fibers in Table 1 below.

The drawbacks of boron fibers are: a) a tendency to further grain growth at high temperature, b) the high reactivity of boron with many metals and the formation of brittle intermetallic phases and c) high cost. The reactivity problem can be partially remedied with the use of intermediate coatings (see Section 2.5 below).

Applications: Boron fibers are used as unidirectional reinforcement for epoxy composites, in the form of prepreg tape. The material is used extensively mostly in fixed and rotary wing military aircraft for horizontal and vertical stabilizers, rudders, longerons, wing doublers and rotors. They are also used in sporting goods. Another major application is as reinforcement for metal matrix composites, in the form of an array of fibers pressed between metal foils, the metal being aluminum in most applications.

TABLE 1

Selected Properties of B and SiC Fibers

	CVD Boron*	CVD SiC*	Sol-gel SiC **
Fiber diameter, μm	100	150	10-20
Density, g/cm^3	2.5	3.04	2.57
Tensile Strength, GPA	3.6	3.4	2.4-3.2
Tensile Modulus, GPA	400	400	165-200
Thermal expansion, ppm/°C	4.5	1.5	-

* Properties from data sheets, Textron Inc.
Lowell, MA 01851

** Properties from data sheets, Nippon Carbon Co.
Tokyo, Japan

Boron has high neutron absorption and the boron-aluminum composites are being investigated for nuclear applications. Single-ply boron-epoxy composites have microwave polarization properties with potential commercial and other applications in antenna and radome designs (2).

2.3 The CVD of Silicon Carbide Fibers

CVD silicon carbide fibers are a recent development with promising potential which may take over some of the applications of CVD boron fibers or other refractory fibers, providing that the production cost can be reduced.

Processing: SiC fibers are produced commercially by the reaction of silane and a hydrocarbon in a tubular glass reactor as shown in Figure 1. The substrate is a carbon monofilament pre-coated with a 1 μm layer of pyrolytic graphite to insure a smooth deposition surface and a constant resistivity (6). Other precursors such as $SiCl_4$, CH_3SiCl_3 are also being investigated (7).

Properties: The properties of CVD SiC fibers are shown in Table 1. They are similar to those of CVD boron fibers except that SiC is more refractory and less reactive than boron. CVD SiC fibers retain much of their mechanical properties when exposed to high temperature in air up to 800°C for as long as one hour as shown in Figure 2 (8). SiC reacts with some metals such as titanium in which case a diffusion barrier is applied to the fiber (see Section 2.5 below).

Applications: Most applications of CVD SiC fibers are still in the development stage. They include the following:

• Reinforcement for metal-matrix composites with such metals as titanium, titanium aluminide, aluminum, magnesium and copper. Applications are found mostly in advanced aerospace programs and include fan blades, drive shafts and other components.

• Reinforcement for ceramic and polymer composites

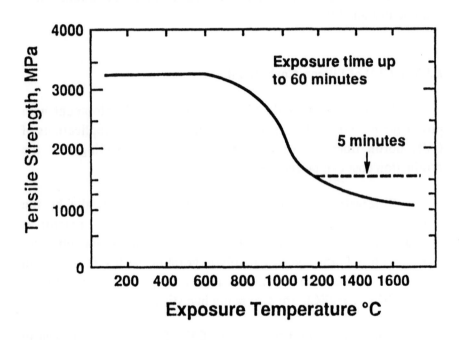

Figure 2. Tensile Strength of CVD SiC Fibers as a Function of Temperature

2.4 Other CVD Fiber Materials

Other materials besides boron and SiC with potential for CVD fiber production are being investigated. They include B_4C, TiC and TiB_2 deposited on a heated core filament, generally by the hydrogen reduction of the chlorides (7).

Typical properties of the resulting experimental fibers are

shown in Table 2. Fiber diameter varies from 20 to 200 μm.

Table 2

Selected Properties of CVD Fibers (7)

	B_4C	TiC	TiB_2
Melting Point,°C	2450	3180	2980
Deposition Temperature, °C	1300	1400	1250
Thermal Expansion, ppm/°C	5.0	7.6	7.9
Tensile Strength,GPa	2.07-2.76	1.38-2.07	1.38-2.07
Modulus, GPa	172-241	172-241	138-207

2.5 CVD Coatings for Fibers

Inorganic fibers, whether made by CVD, sol-gel or other processes, are often prone to react with the metal or ceramic matrices during processing into composites or during operation at high temperature. Such reactions produce intermetallics or other compounds which may considerably degrade the properties of the fiber. In such cases, it is necessary to apply a coating on the fiber that acts as a diffusion barrier and prevents this diffusional reaction. The most common technique for applying such coatings is CVD (9).

A typical fiber coating apparatus is shown schematically in Figure 3. Common diffusion barrier materials are pyrolytic carbon, BN, TiN, TiB_2, SiC and rare earth sulfides (10,11,12).

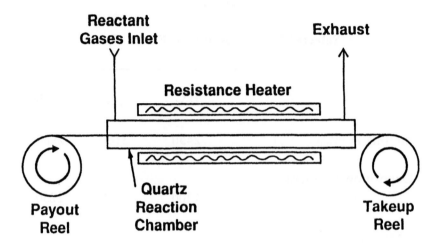

Figure 3. Schematic Diagram of CVD Fiber Coating Apparatus

2.6 Whiskers

Whiskers are short fibers, usually single crystals, with an aspect ratio of 10/1 or greater. They have very high strength and are used as a random reinforcement for ceramics or metals. CVD is a common method of growing whiskers (13). The following refractory CVD whiskers have recently been produced.

• Titanium diboride whiskers by the hydrogen co- reduction of $TiCl_4$ and BBr_3 in the presence of a platinum catalyst. These whiskers are 20-30 nm thick and 12-13 μm in length (14)

• Titanium carbide whiskers from $TiCl_4$, CH_4 and H_2 in the presence of a nickel catalyst at 1220-1450°C (15)

• Hafnium carbide and hafnium nitride whiskers from $HfCl_4$, CH_4 or N_2, and H_2 in the presence of a nickel catalyst at 1000-1450°C (16)

3.0 CVD IN POWDER APPLICATIONS

3.1 CVD Ceramic Powders

The preparation of ceramic powders by CVD is a promising and growing field. These powders are used as precursors for the processing of advanced ceramics, or directly as powder for such applications as abrasives.

The general trend in ceramic technology is toward better batch-to-batch reproducibility of properties, improved physical or chemical properties and lower processing temperature. New materials such as silicon nitride, silicon carbide, and partially stabilized zirconia have strength which is achieved by using pre-cursor powders with great purity, small particle size, uniformity, and freedom from aggregation. These powders are now used to produce a wide variety of ceramics such as structural components, catalyst supports, pigments, electro-optical and magnetic devices, drug-delivery carriers and many others.

In the traditional ceramic powder preparation, the raw material is processed through a series of steps which are typically: a) heating or calcination to obtain a thermodynamically stable material, b) grinding and milling to a suitable particle size and distribution, c) dispersion to control the agglomeration process. Satisfactory for most purposes, these powders do not generally meet the exacting requirements of advanced ceramic applications.

The consensus is that higher quality ceramic powders can be

obtained by chemical means such as combustion synthesis, sol-gel or CVD (17). In this Section, the process of synthesizing powders by CVD and their major applications are reviewed.

CVD Process for Powder Production: If the temperature and supersaturation are sufficiently high in a CVD reaction, the product is primarily powder precipitated from the gas phase (see Chapter Two, Section 4.1). Such powders have very few impurities provided that the CVD precursors are carefully purified. Their small diameter and great uniformity are important factors in the production of high quality hot-pressed or sintered ceramic bodies with good mechanical and electrical properties. In addition, the sintering temperatures required for CVD powders are lower than those for conventional powders.

CVD Ceramic Powder Materials: The following powders are made experimentally or on a production basis.

• Beta SiC powder is produced from methyl-trichlorosilane (MTS) in the presence of hydrogen in an argon plasma. This powder is pressureless sintered to a maximum density of 97% of theoretical (16). Submicron SiC powder is also prepared by the gaseous thermal decomposition of tetramethylsilane, $Si(CH_3)_4$, in a flow-through reactor between 850 and 1500°C (18). Very small powders of 10 nm diam. are produced by using acetylene and silane as reactants (19).

• An amorphous silicon nitride powder is produced from silicon halides and ammonia at high temperature (20). The powder can also be produced by using the same reaction at 1000°C in an RF plasma with a mean particle size of 0.05-0.1 μm (21). Si_3N_4 powder is also produced by laser CVD from halogenated silanes and ammonia with an inert sensitizer such as SF_6. A similar process is used to produce silicon powder (22).

• Other CVD ceramic powders of submicron size include

aluminum nitride produced from aluminum alkyl (23), magnesium oxide produced from Mg vapor and O_2 at 800°C (24) and tungsten carbide (25).

Applications: CVD ceramic powders such as SiC and Si_3N_4 are used to produce ceramic bodies for a wide variety of applications, either experimentally or in production. These include structural applications in high temperature or corrosive environments where metals are not suitable, in such areas as reciprocating engines, gas turbines, turbochargers, bearings, machinery and process equipment.

Another area is superconductivity where powders of $YBa_2Cu_3O_7$ produced by CVD are used in the fabrication of sputtering targets and may also be used in plasma spraying for the fabrication of superconductor wires and coatings (see Chapter Seven, Section 5.12).

Rapidly growing in importance is the production of high thermal conductivity heat sink material for integrated circuits from CVD AlN powders (see Chapter Eight, Sections 7 and 8).

How much of an impact CVD ceramic powders will have on ceramic technology is not clear at this time since applications have yet to reach the commercial stage in any significant manner. Yet the success of a number of development programs strongly suggests that CVD will soon become a major technology in this field. This is already evident in Japan which is the recognized leader in advanced ceramics.

3.2 CVD Metal Powders

The processing of metals by powder metallurgy has unique advantages and the resulting products have excellent properties

(26). Several metal powders used in powder metallurgy are now produced by CVD. These include:

• Iron, nickel, cobalt, molybdenum and tungsten powders produced by the pyrolysis of the metal carbonyls (see Chapter Three, Section 4) (27).

• Tungsten, produced from WF_6 to a purity in excess of 99.9995% (28).

3.3 Coated Powders

Metallic and ceramic powders as small as 5 μm in diameter can be coated by CVD by the fluidized bed technique (see Chapter four, Section 10). Particles can be coated free of agglomeration with good uniformity. The most common coating materials are nickel and iron which are obtained by the decomposition of the respective carbonyl (see Chapter Three, Section 4). These metals are used to coat tungsten particles to promote sintering in powder metallurgy applications (29,30). Sintering time, sintering temperature and grain growth are reduced, contamination is lessened and properties are improved.

A major problem in fluidized-bed coating is particle agglomeration. As the particle size is decreased, it becomes more difficult to avoid agglomeration and to coat the particles individually. In general, the tendency to agglomerate increases with decreased particle size, decreased density and with increased temperature. It is also a function of the materials (substrate and coating), the particle shape (acicular vs. equiaxed), and the coating atmosphere (presence or absence of molecular species). The composition, configuration and flow dynamics of the fluidizing atmosphere can be adjusted within limits to promote unagglomerated fluidization.

Other potential applications are ceramic powders coated with their sintering aids, zirconia coated with yttria stabilizer, tungsten carbide coated with cobalt or nickel, alumina abrasive powders coated with a relatively brittle second phase such as $MgAl_2O_4$ and plasma spray powders without the segregation of alloying elements.

4.0 CVD IN MONOLITHIC AND COMPOSITE APPLICATIONS

Some of the earliest applications of CVD were monolithic (or free-standing) structures. A tungsten tube made by CVD was reported as early as 1932 (13). The applications of CVD in this area are now well established and many are in substantial commercial production. In this Section, these applications are reviewed.

4.1 Graphite, Carbon-Carbon and Boron Nitride CVD Structures

Pyrolytic Graphite: CVD graphite (also known as pyrolytic graphite) has a strong structural anisotropy and a corresponding anisotropy in its properties (see Chapter Six, Section 3). It is used in applications which require high strength and corrosion resistance at high temperature in non oxidizing environments. Such applications include crucibles, semiconductor wafer handling equipment, electrodes, reaction vessels for gas-phase epitaxy and rocket nozzles.

Carbon-Carbon: Carbon-carbon composites consist of a carbon fiber substrate, either in the form of woven fabric or random felt, and a carbon matrix (31). The matrix is produced by the pyrolysis of a polymer, usually phenolic, reinfiltrated by CVD or by chemical vapor infiltration (CVI) alone (see Chapter Four, Section

9). The material has outstanding high temperature mechanical properties and excellent chemical resistance except to oxidation. It is a high cost material and its uses are mostly limited to aerospace at the present time. Some applications are as follows.

• Carbon-carbon has antigalling characteristic with the ability to slide against itself with very little wear, making it ideal for high-performance brakes. Such brakes are probably the major application of carbon-carbon and are found on most military aircraft, on the space shuttle. They are being introduced on civilian aircraft and racing cars (31).

• The space shuttle and other aerospace vehicles use carbon-carbon extensively in nose cap, leading edges, structural panels and other components.

• Experimental gas turbine rotors have demonstrated the potential of carbon-carbon provided that the oxidation problem can be solved.

• Other applications include missile nozzles and nose cones, aerospace fasteners, heating elements and glass bottle machinery components.

Boron Nitride: Pyrolytic boron nitride produced by CVD is very inert chemically and, like pyrolytic graphite, has highly anisotropic properties (see Chapter Seven, Section 4.2). For instance, its thermal conductivity in the a-b direction is 40 times that of the c direction. It is much more oxidation resistant than graphite. As such, boron nitride is an ideal material for high temperature containers and is manufactured on a large scale for crucibles for aluminum evaporation and for molecular beam epitaxy, for Czochralski crystal growth vessels of III-V and II-VI compounds and for furnace hardware. It is also produced as an insulating substrate in ribbon heaters in combination with a pyrolytic graphite resistance heating element (33,34).

4.2 Monolithic Metallic Structures

Many refractory metals are difficult to fabricate by conventional metallurgical processes because of their very high melting point. They are usually hard and brittle due to the presence of impurities which makes them very difficult to machine. With CVD, it is possible to fabricate these metals to net shape without further need to machine them. In addition, CVD produces metals which can be made extremely pure (greater than 99.9%) with very little carbon, oxygen or other contaminants that are the major cause of embrittlement in powder metallurgy processing of refractory metals.

The refractory metals for which CVD is commonly used to produce free-standing shapes are tungsten, niobium, rhenium, tantalum, molybdenum and nickel (13, 35). The CVD of these metals is described in Chapter Five. Shapes presently produced include rods, tubes, crucibles, manifolds, ordnance items, nozzles and thrust chambers. They are usually deposited on a disposable mandrel of copper, molybdenum or graphite which is subsequently machined off or removed chemically by etching.

4.3 CVD Ceramic Composites

The major drawback of ceramics is their intrinsic brittleness. For example, most metals have a fracture toughness forty times greater than conventional ceramics and glasses. This difference is related on the atomic level to the strong hybrid ionic-covalent bonds of ceramics. These high strength bonds prevent the deformation such as occurs in ductile metals. Since ceramics cannot deform, they fail catastrophically. Applied stresses tend to concentrate at the sites of flaws such as voids, notches, scratches or chemical impurities at grain interfaces.

Ceramic composites, which use ceramic fiber or whisker reinforcement in a ceramic matrix, are less susceptible to brittle failure since the reinforcement intercepts, deflects and slows crack propagation. At the same time, the load is transferred from the matrix to the fibers to be distributed uniformly. Such composites are characterized by low density, generally good thermal stability and corrosion resistance. Their theoretical strength is very high, but has yet to be completly achieved because of the non-ductile behavior of the matrix which results in pronounced notch sensitivity and leads to catastrophic failure.

The development of suitable intermediate coatings between fiber and matrix, deposited by CVD, which allow a certain degree of load transfer to the fiber, has improved mechanical properties to some degree.

However, exposure to high temperature (i.e.> 1000°C) causes diffusion and chemical bonding across the interface and brittle failure is still dominant. A great deal of experimental work is under way to remedy these conditions.

The reinforcing fibers are usually CVD SiC (see Section 2.3 above) or modified aluminum oxide. A common matrix material is SiC deposited by chemical vapor infiltration (CVI), a process which is described in Chapter Four, Section 9. The CVD reaction is based on the decomposition of methyl-trichlorosilane at 1200°C. Densities approaching 90% are achieved (36). Another common matrix material is Si_3N_4 which is deposited by isothermal CVI using the reaction of ammonia and silicon tetrachloride in hydrogen at 1100-1300°C and a total pressure of 5 torr (37). The energy of fracture of such a composite is considerably higher than that of unreinforced hot-pressed silicon nitride.

To this date, the fabrication of structural ceramic composites has been limited to prototypes mostly in high-cost, high-performance aerospace applications such as missile guidance fins, hypersonic fuselage skins, inner flaps and rocket nozzles.

If the composite is left only partially densified, it can be used as a filter for high temperature filtering systems with high collection efficiency as required in direct coal-fired gas and steam turbines (38). Similar systems are considered for particulate filtering in Diesel engines by a carbon foam coated with silicon carbide by CVI.

REFERENCES

1. Private Communication, Marketing Data Bank, *G.A.M.I.*, Gorham, ME 04038 (1991)

2. Buck, M.E., Advanced Fibers for Advanced Composites, *Advanced Materials and Processes*, 61-65, (Sept. 1987)

3. Mah, T., Mendiratta, M., Katz, A. and Mazodiyasni, K., Recent Developments in Fiber-Reinforced High Temperature Ceramic Composites, *Ceramic Bull.*, 66(2):304-317 (1987)

4. *Ceramic Bulletin*, 66(3):424 (1987)

5. Krukonis, V., in *Boron and Refractory Borides*, (V. Matkovich, ed.) 517-540, Springer Verlag, New York (1977)

6. Sheppard, L.M., Progress in Composites Processing, *Ceramic Bulletin*, 69(4):666-673 (1990)

7. Gatica, J.E. and Hlavacek, V., Laboratory for Ceramic and Reaction Engineering: A Cross-Disciplinary Approach, *Ceramic Bulletin*, 69(8):1311-1318 (1990)

8. AVCO Silicon Carbide Fiber, Technical Brochure, *Textron*, Lowell, MA 01851 (1990)

9. Cranmer, D.C., Fiber Coating and Characterization, *Ceramic Bulletin*, 68(2): 415-419 (1989)

10. Bouix, J., Vincent, H., Boubehira, M, and Viala, J.C., Titanium Diboride-Coated Boron Fibre for Aluminum Matrix Composites, *J. Less Common Metals*, 117(1-2):83- 89 (Mar. 1986)

11. Kerans, R.J., Hay, R., Pagano, N. and Parthasarathy, T., The Role of the Fiber-Matrix Interface in Ceramic Composites, *Ceramic Bulletin*, 68(2): 429-442 (1989)

12. Flahaut, J., *Progress in Science and Technology of Rare Earths*, Vol.3 (Leroy Eyring, Ed.), 209-283, Pergamon Press (1968)

13. Blocher, J.M., Jr., Structural Shapes, in *Vapor Deposition*, (C. Powell et al, Eds.) 651-692, John Wiley & Sons, New York (1966)

14. High Srength Titanium Boride Whiskers, *Japanese New Materials High Performance Ceramics*, Jan. 1989

15. Wokulski, Z. and Wokulska, K., On the Growth and Morphology of TiC_x Whiskers, *J. Cryst. Growth*, 62(2):439-446 (July 1983)

16. Futamoto, M., Yuito, I. and Kawabe, U., Hafnium Carbide and Nitride Whisker Growth by Chemical Vapor Deposition, *J. Cryst. Growth* 61(1):69-74 (Jan. Feb. 1983)

17. *Ceramic Technology Newsletter*, Apr. Jun. 1990, Oak Ridge National Lab., Oak Ridge, TN 37831-6050

18. Wu, H.D. and Ready, D.W., Silicon Carbide Powders by Gaseous Pyrolysis of Tetramethylsilane, in *Silicon Carbide 87, Ceramic Transactions*, Vol. 2:35-46 (1987)

19. High Purity Ultrafine SiC Powder, in *Advanced Ceramic Reports,* Vol. 3(9):2 (1988)

20. Schwier, G., On the Preparation of Fine Silicon Nitride Powders, in *Progress in Nitrogen Ceramics*, (F. Riley, ed.) 157-156, Martinus Nijhoff, Boston, MA (1983)

21. Hussain, T. and Ibberson, V., Synthesis of Ultrafine Silicon Nitride in an RF Plasma Reactor, in *Advances in Low-Temperature Plasma Chemistry, Technology, Applications*, Vol.2:71-77, (H. Boenig, Ed.), Technomic, Lancaster (1984)

22. Bauer, R., Smulders, R., Geus, E., van der Put, J. and Schoomman, J., Laser Vapor Phase Synthesis of Submicron Silicon and Silicon Nitride Powders from Halogenated Silanes, *Ceram. Eng. Sci. Proc.*, 9(7-8):949-956 (1988)

23. *Japan Chemical Week*, p.5, (July 13, 1989)

24. Kato, A., Study of Powder Preparation in Japan, *Ceramic Bulletin*, 66(4):647-650 (1987)

25. Hojo, J., Gono, R. and Kato, A., The Sinterability of Ultrafine WC Powders by a CVD Method, *J. Mat. Sci.*, 15(9):2335-2344 (1980)

26. German, R.M., Powder Fabrication, in *Powder Metallurgy Science*, Metal Powder Industries Federation, Princeton, NJ (1984)

27. Powell, C.F., Chemically Deposited Metals, in *Vapor Deposition*, (C.F. Powell et al, Eds.), John Wiley & Sons, New York (1966)

28. *Japanese New Materials Advanced Alloys and Metals*, Feb. 1989

29. Williams, B.E., Stiglich, J. and Kaplan, R., CVD Coated Powder Composites, Part I: Powder Processing and Characterization, Ultramet, in *Proc. TMS Ann. Meeting, New Orleans, LA (Feb. 1991)*

30. Mulligan, S. and Dowding, R.J., Sinterability of Tungsten Powder CVD Coated with Nickel and Iron, Tech. Rep. TR-90-56, *U.S. Army Mat. Tech. Lab.*, Watertown, MA (1990)

31. Buckley, J.D., Carbon-Carbon, an Overview, *Ceramic Bulletin*, 67(2):364-368 (1988)

32. Awasthi, S. and Wood, J., Carbon/Carbon Composites for Aircraft Brakes, *Ceram. Eng. Sci. Proc.*, 9(7-8):553-560 (1988)

33. Lelonis, D.A., Boron Nitride Tackles the Tough Ones, *Ceramic Industry*, 57-60 (April 1989)

34. Iwasa, M. and Ide, M., Properties and Cycle Life of Pyrolytic BN Crucibles, *Proc. 10th. Int. Conf. on CVD* (G. Cullen, Ed.), 1098-1105, Electrochem. Soc., Pennington, NJ 08534 (1987)

35. Kaplan, R. and Tuffias, R., Fabrication of Free Standing Shapes by Chemical Vapor Deposition, *Proc. of AES Int. Symp. on Electroforming/Deposition Forming,* Los Angeles, CA (Mar. 1983)

36. Veltri, R.D., Condit, D. and Galasso, F., Chemical Vapor Deposited SiC Matrix Composites, *J. Amer. Ceram. Soc.*, 72(3):478-480 (1989)

37. Foulds, W., LeCostaouec, J., Landry, C. and DiPietro, S., Tough Silicon Nitride Matrix Composites Using Textron Silicon Carbide Monofilaments, *Ceram. Eng. Sci. Proc.*, 10(9-10):1083-1099 (1989)

38. Stinton, D.P. and Lowden, R.A., Fabrication of Fiber-Reinforced Hot-Gas Filters by CVD Techniques, *Ceram. Eng. Sci. Proc.*, 9(9-10):1233-1244 (1988)

APPENDIX

ALTERNATIVE PROCESSES FOR THIN-FILM DEPOSITION AND SURFACE MODIFICATION

1.0 ION IMPLANTATION

In the ion-implantation process, ions are produced in a source, then accelerated at high voltage (several keV or meV) and impacted and implanted in the surface to be modified (1). This modification is obtained without the interface characteristic of coatings. The major limitation of ion implantation is the shallow depth of the treatment, from 0.1 to 0.5 μm, depending on the accelerating voltage and the mass of the ion. The equipment and the process are expensive, size is limited and complex shapes are difficult to implant because of line-of-sight limitations.

Ion implantation is a major process in the fabrication of semiconductors. It is used to implant dopants such as arsenic, phosphorus or boron to modify the electrical properties of the semiconductor substrate. Two interesting developments are the cathodic arc and the metal ion source for metal implantation.

2.0 SOL-GEL

Sol-gel processing or, more broadly speaking, the processing of ceramics and glasses from chemical precursors, is an emerging technology with a great potential especially in coatings and fiber production. At present, the development of sol-gel is still predominantly in the laboratory, yet it has spurred a great deal of interest in high-technology industries such as semiconductors, opto-electronics, optics, and structural ceramics. Sol-gel is already used in a sizable range of applications particularly in thin films. These applications include the spin-on-glass (SOG) process which is used in semiconductor fabrication and the coating of architectural glass to control optical properties.

The sol-gel process uses a liquid reactive precursor material that is converted to the final product by chemical and thermal means. This precursor is prepared to form a colloidal suspension or solution (sol) which goes through a gelling stage (gel) followed by drying and consolidation. The process requires only moderate temperatures, in many cases less than half the conventional glass or ceramics processing temperatures. It also permits very close control of the composition and structure of the deposit on the molecular level and makes possible the production of near-net shapes. These are very important advantages.

A number of processing problems still need to be solved. The need for better control of shrinkage and cracking is critical, and the cost of precursor materials is still high, although it is likely to come down as applications are developed and consumption increases (2).

3.0 PHYSICAL VAPOR DEPOSITION (PVD)

Like CVD, PVD is a vapor deposition process where the deposition species are atoms or molecules or a combination of these. The distinction between these processes is that, in CVD, deposition occurs by chemical reaction, whereas in PVD, deposition is by condensation. CVD is usually endothermic at the substrate and PVD is exothermic.

The source of the PVD species may be evaporation or sputtering. An important recent trend is the tendency for the two processes, CVD and PVD, to merge. CVD for instance, now makes extensive use of plasma (a physical phenomenon) and conversely, reactive evaporation and reactive sputtering occur in a chemical environment. CVD and PVD reactors are now combined in one single piece of equipment in new semiconductor processing operations and the difference between the two processes becomes blurred. PVD processes are described as follows.

3.1 Evaporation

Evaporation is the simplest PVD process. It is also known as vacuum evaporation. The coating material is heated in a vacuum above its boiling point, sending atoms or molecules in a straight line to the substrate, where these condense to form a thin film. A vacuum of 10-5 to 10-6 Torr is preferable. Heating of the coating material is accomplished by direct resistance, laser beam, arc discharge and increasingly by electron beam. Very high deposition rates are now possible that may reach 75 μm/min. With reactive evaporation, it is possible to deposit compounds such as carbides, nitrides and oxides. The drawbacks of evaporation are generally poor adhesion of the coating and difficulty in coating complex shapes. Evaporation is used extensively in optical and decorative applications, mostly for the deposition of metals. Very large, continuous coating equipment is now available (3).

Evaporation is facing strong competition from other thin film processes such as sputtering and CVD and is generally on the decline. Two techniques based on evaporation may reverse this trend. One is based on a pulsed excimer laser which acts as a flash vaporizer and can evaporate compounds with elements having widely different vapor pressures and complex composition so that the original composition is retained in the deposited film.

The other technique is molecular beam epitaxy (MBE). MBE produces extremely pure and very thin films with abrupt composition changes. Deposition rate however is very slow. It is used for the production of very exacting (and expensive) microwave and opto-electronic devices (4).

3.2 Sputtering

Sputtering and CVD are now the dominant deposition technologies. It is only in the last 25 years or so that sputtering has emerged as a large-scale production process, particularly in the semiconductor and tool-coating industries (5).

The principle of sputtering is relatively simple. A source (the cathode, also called the target) is bombarded in a high vacuum with gas ions (usually argon) that have been accelerated by a high voltage, producing a glow discharge or plasma. Atoms from the target are ejected by momentum transfer and move across the vacuum chamber to be deposited on the substrate to be coated. Very low initial pressure is required (10^{-7} Torr) to remove impurities from the system. The development of a magnetically enhanced cathode or "Magnetron" has considerably expanded the potential of sputtering. The Magnetron sends the electron into spiral paths to increase collision frequency and enhance ionization.

Reactive sputtering occurs when the species are sputtered in a reactive atmosphere such as nitrogen or methane. The process is now used extensively for the deposition of nitrides and carbides. Sputtering can produce a wide variety of coatings with excellent adhesion and good composition control, without the high temperature requirements of CVD. Its major disadvantages are its low deposition rate and its line-of-sight characteristic. The latter can be overcome to some extent by operating at higher pressure, but at some sacrifice in deposition rate.

3.3 Ion Plating

Ion plating is an hybrid process in which the coating material is vaporized in a manner similar to evaporation but passes through a gaseous glow discharge or plasma, as it travels toward the substrate, thus ionizing many of the atoms. The plasma is obtained by biasing the substrate to a high negative potential (5kV) at low pressure. The constant ion bombardment of the substrate sputters off some of the surface atoms which results in improved adhesion and reduced impurities. Step coverage is also improved (6).

The process is used in the production of optical thin films, in wear- and erosion-resistant applications and in some electrical contact designs.

REFERENCES

1. Picraux, S. and Pope, L., Tailored Surface Modifications by Ion Implantation and Laser Treatement, *Science*, 226-267 (Nov. 1984)

2. Brinker, C.J. and Scherer, G.W., *Sol-Gel Science*, Academic Press, San Diego, CA (1990)

3. Bunshah, R.F. and Deshpandey, C.V., Evaporation Processes, *MRS Bulletin*, 33-39, (Dec. 1988)

4. Moustakas, T., Molecular Beam Epitaxy: Thin Film Growth and Surface Studies, *MRS Bulletin*, 29-34, (Nov. 1988)

5. Westwood, W.D., Sputter Deposition Processes, *MRS Bulletin*, 46-51 (Dec. 1988)

6. Rossnagel, S.M. and Cuomo, J.J., Ion-Beam Deposition Film Modification and Synthesis, *MRS Bulletin*, 40-45, (Dec. 1988)

INDEX

414

Printed and bound by CPI Group (UK) Ltd, Croydon, CR0 4YY

08/05/2025

01864823-0002